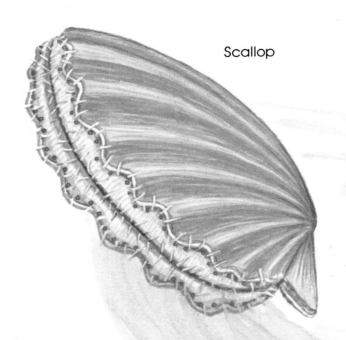
Scallop

Mysterious Seas

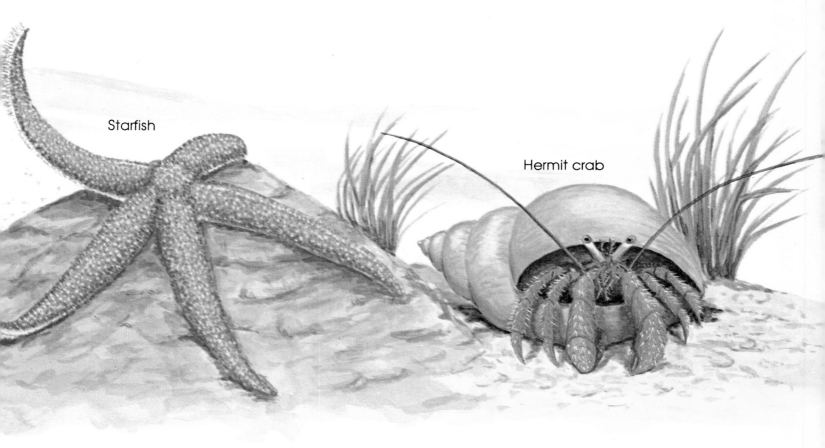

By Mary Elting
Illustrated by Fiona Reid

Publishers • GROSSET & DUNLAP • New York

How to Pronounce some of the Names in this Book

Anemone—uh-NEM-oh-nee
Ciliate—SILLY-ate
Coccolith—COCK-oh-lith
Ctenophore—TEEN-oh-fore
Diatom—DIE-ah-tome
Flagellate—FLADGE-uh-late
Foraminifera—FOR-am-un-IFF-er-ah
Janthina—YAN-thin-ah
Kraken—KRAH-ken
Medusa—Muh-DOO-sah
Megamouth—MEG-ah-MOUTH
Noctiluca—NOCK-tee-LOO-kah
Nudibranch—NOOD-ee-BRANK
Oikopleura—OY-koh-PLOOR-ah
Polyp—POLL-IP
Pyrosoma—PIE-roe-SO-mah
Radiolaria—RAY-dee-oh-LARE-ee-a
Sargassum—Sar-GAS-um
Tsunami—Tsoo-NAM-ee
Velella—Vel-ELL-ah
Wrasse—RASS

Text copyright © 1983 by Mary Elting. Illustrations copyright © 1983 by Fiona Reid. All rights reserved. Printed in the United States of America. Library of Congress Catalog Card Number: 83-47614. ISBN: 0-448-18960-7. Published simultaneously in Canada.

Contents

The Secret Supermarket	8
The Impossible Worms and Shrimps	16
The Case of the Walking Flower	20
The Case of the Hummingbird and the Rosebush	23
Some Strange Equipment	26
Senses and Extra Senses	33
Things That Go Beep in the Deep	39
Babies in Disguise	46
Eerie Lights in the Dark	50
Safety Devices	56
The Monster and the Giant	61
Strange Rivers and Perilous Waves	65
Solving Mysteries—Preserving Life	70
Glossary	76
Index	77

Author's Acknowledgments

The wonders you will find in this book were discovered by scientists, many of whose books and articles I have read with excitement and gratitude. For help in finding answers to questions, I want to thank the people in the library of the National Oceanic and Atmospheric Administration in Boulder, Colorado. For suggestions and encouragement I thank Raphael Folsom, who will surely grow up to be a marine biologist. Special thanks go to the Harvard Museum of Comparative Zoology and to its Director, Dr. James J. McCarthy, who read the manuscript with an eye to making it as accurate as possible. If errors remain, they are my responsibility, not his.

Illustrator's Acknowledgments

I would like to thank Guy Tudor for his artistic advice and help with reference material, and my friends in Old Field and the Marine Science Center at Stony Brook for their encouragement and helpful criticism.

Marlin chasing grey mullet

The Secret Supermarket

Big fish eat little fish, and little fish eat smaller ones. But what do the littlest ones eat? Seaweeds? Good question, but it took a very long time to find the answer.

For thousands of years people sailed the oceans and caught fish. They knew about big fish and little fish. They saw plants growing in the sea and found some creatures nibbling on plants. Everybody supposed that sea life must be very much like life on land. That is, the meat-eaters fed on the creatures that ate plants.

But this was just a guess. Nobody had really looked for evidence. In some mysterious way the sea provided plenty of food for all its creatures, but no one could see how.

The important word turned out to be "see."

It was not until fairly good microscopes had been invented that scientists and professors began to pay more attention to sea water. They had known for many years that water in ponds held living creatures too small to see without a microscope. Then, about 150 years ago, some students at a university in Denmark got excited about looking for small things in the sea.

To collect their invisible specimens they made a clever invention. Ordinary nets, of course, would not do. The holes in them were too big. But their professor found just the right thing at a grain mill in the country. The miller sifted ground-up wheat through silk cloth so fine that only the powdery flour could get through. Nets made of this silk would let water out, but microscopic living things would be caught against the cloth.

Each scoop of water brought the students dozens of different specimens to study. Some were clearly animals. But others, called *diatoms*, caused a great argument. Were they animals or not? These tiny specks of life were able to build houses around themselves. And what houses they were! Under the microscope some kinds of diatom looked like glass jewel boxes. Others resembled chains of opals, stars, glassy wheels, or propellers. All of them were pierced with little holes like windows. Inside their thousands of different-shaped shells, the bodies of diatoms seemed to be bits of golden or brownish green jelly. Some contained a drop of oil.

Finally everyone agreed that the diatoms were members of the plant kingdom. Each one of these bits of living jelly was a complete plant. Its body was just one single unit or cell, and it had no leaves, roots or stems. But it contained a bit of green stuff called *chlorophyll* that other plants have, and it could behave like other plants in important ways.

A selection of diatoms, greatly magnified

Scientists began to study diatoms carefully, and what they found was not just a simple little blob of jelly. The outside wall of a diatom cell can take a mineral that is dissolved in sea water and turn that mineral into its glassy outside shell. This is the very same mineral that bottle factories make into glass bottles.

Another substance that is dissolved in sea water is carbon dioxide—the same gas that makes soda water fizz. A diatom needs this gas in order to grow. It also needs tiny amounts of certain minerals and even vitamins that are in the sea. With the help of sunlight and its chlorophyll, the diatom uses carbon dioxide, water, and vitamins and minerals to grow in much the same way other plants grow.

It is hard to imagine a diatom at work. How can it use the sun's energy to turn those substances into nourishing food, complete with its glass shell to hold the food? Scientists still don't completely understand all the diatom's secrets. They do know that many separate chemical steps are needed to make the one-celled plant. The changes from one step to another happen very fast, and so diatoms grow much faster than land plants. There can be two million of them in a cup of sea water when they are growing best in springtime.

Diatoms and some other microscopic plants make up the great meadows of the sea. They provide the food for most of the plant-eating fish and many other sea creatures.

But plants were not the only bits of life that students found under their microscopes. They discovered many kinds of animal as well. Like animals on land, they cannot make their food the way plants do. So some of them feed on the microscopic plants. Some others use material that dissolves in the water when dead plants and animals decay. Still others use leftovers—the particles of droppings that remain after food has been digested and passed out of the bodies of larger sea creatures.

Microscopic sea animals come in many astonishing shapes. One scientist collected 3,508 different new animals called *radiolaria* when he made a voyage around the world on a ship called the *Challenger*. The beautiful skeletons of radiolaria are glassy spikes made of the same mineral that the diatoms use in forming their shells. The animals' bodies are wrapped around the spikes, and they throw out sticky threads to capture the microscopic plants that are their food. Some kinds of radiolaria even grow their own gardens of these plants inside their tiny bodies.

On that same voyage around the world, another scientist decided to take the ship's rowboat out one day when the ocean was very calm. As he sat studying the water, a speck of red caught his eye. Carefully, using a teaspoon he happened to have with him, he scooped it up. Back on the ship, he looked at the red speck under his microscope. And he saw something that no one had ever captured before. Unlike the radiolaria, its soft little red body clustered around a skeleton made of a chalky substance like that of a clam shell. Because the skeleton is full of little windows, scientists gave this group of creatures the name *foraminifera* which means just that—full of windows. People who study them call them forams for short.

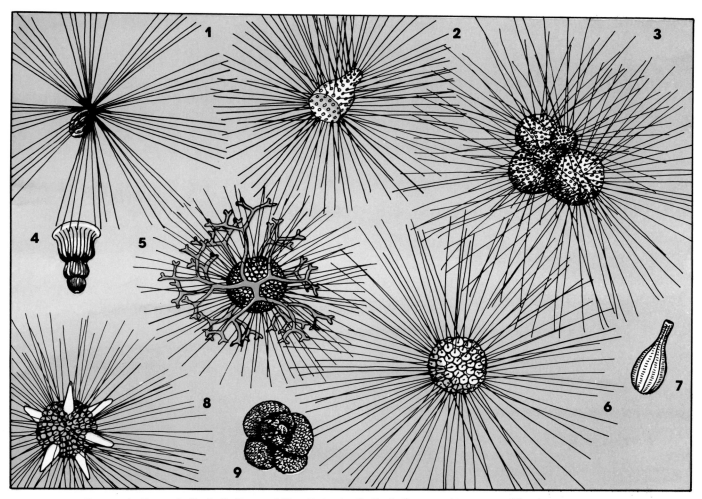

Foraminifera 1, 3, 4, 7, 9, and Radiolaria 2, 5, 6, 8, greatly magnified

Diatoms, radiolaria, forams, along with bacteria and hundreds of other small forms of life make up the secret supermarket of the sea. The diatoms are food for little shrimplike animals. Some of these are the size of a pinhead, some about as big as a grain of rice, others a little larger. One of the larger kind, called *krill*, can gobble up 120,000 diatoms in a day.

This immense crowd of little animals and plants drifts about in water where larger creatures find them and eat them. The scientific name for all of them together is *plankton*, which means drifters or wanderers. Without the plankton there could not be food enough for everything else in the sea.

Here is why:

Ten tons of microscopic plants are needed to produce one ton of animals in the plankton.

Ten tons of animals in the plankton are needed to produce one ton of small fish.

Ten tons of small fish are needed to produce one ton of big fish.

This means that 2 million pounds of tiny plants must grow in the sea meadows if just four bluefin tuna, are to reach a weight of 500 pounds each.

Even more amazing, some kinds of whale depend directly on the plankton. Their main food is the little shrimplike krill.

Marvels to be found in the plankton are endless. Many microscopic bits of jelly seem to have tails like whips. Scientists call the whiplike organs *flagella*. One plankton animal has two flagella that spin its round little body through the water like a top. Its scientific name is *ctenophore* but most people call it a sea gooseberry. Another transparent, barrel-shaped animal, about an inch long, has a curious way of multiplying. One by one, new little barrels form inside its body. Then, strung together like the tail of a kite, they stream out, still attached to their parent. These creatures are called *salps*.

One sea mystery puzzled the experts for a while. They knew that billions of very tiny shells covered the ocean floor in places, but no one knew which creature the shells came from. These same shells were also found, along with foram skeletons, in chalky white cliffs in England. Long ago, this part of England lay beneath water. Animals lived and died in the water, and their shells and skeletons piled up in layers and layers on the sea bottom. Later, great forces in the earth pushed the shell-covered area high and dry to form the white cliffs.

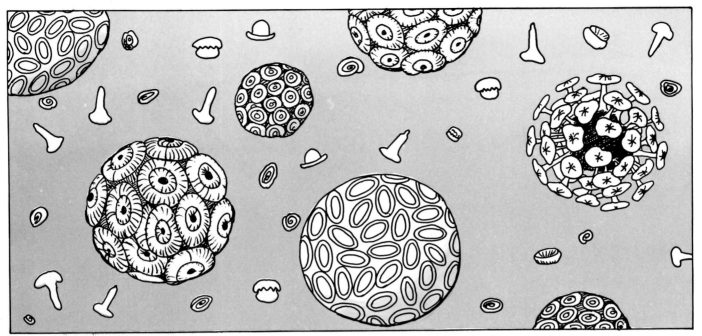

Coccoliths, greatly magnified

People who collect shells discovered the chalky foram skeletons in the cliffs. But where did the other tiny shells come from? No one had ever seen a living creature inside them. At last, a scientist solved the puzzle. Using a very strong microscope, he discovered that the mysterious bits of white stuff were plates that formed the outside skeletons of some of the sea's smallest inhabitants. Their round bodies were only one five-thousandths of an inch across. Yet they could take a mineral from the water and form a protective coat. These little specks of life were given a big name—*coccolith*, which means something stony and small like a seed.

Did anything smaller than a coccolith exist in the sea? A professor in Germany found the answer, although he wasn't actually looking for it. Instead, he was studying an animal called Oikopleura—Oiky for short. One day he was lucky enough to capture an Oiky alive in its own special house. The animal itself doesn't look like much. Its head and stomach bulge out at one end of a ribbonlike tail, and the whole thing is less than an inch long. The marvellous thing about Oiky is the way it builds its house and captures its food.

From its rather long snout, Oiky spins out a transparent gelatinlike stuff that forms a coat around its body. Then it blows the coat up like a melon-shaped balloon. Two windows, covered with fine screens woven from the same gelatin, are at the back of the house. When Oiky waggles its tail, it makes a current of water flow through the screened windows. The current carries food to two nets that Oiky spins and spreads out in front of its head. After the current brings food to Oiky's nets, the water flows on out through an opening in the front of the balloon house. Oiky has also built a back door into its home. In case of danger, it dashes out of this door, leaving only the balloon for an enemy to eat. Then Oiky goes to work making a new house and new nets like the old ones.

The professor was curious about those nets. What did they catch for Oiky to eat? Using the strongest kind of microscope, the professor found the answer. Oiky's food was something no one had ever seen before—specks of living jelly called *flagellates* that were only one thousandth of a millimeter long—five times smaller than the tiny coccoliths.

Oiky's net was much finer than anything human beings had invented at the time. That was why scientists had never caught this flagellate. Now there are special filters that can trap even these tiniest members of the plankton.

So far scientists and explorers have discovered and named many, many thousands of inhabitants of the oceans. Thirty thousand kinds of fish alone have been identified. Nevertheless, almost every time an expedition comes home from a trip, it brings with it one or more sea creatures that have not yet got a name.

Surrounded by tiny dinoflagellates is Oikey's home with him inside (center), and then swimming freely at right.

The Impossible Worms and Shrimps

The last place scientists expected to find many new creatures was a mile and a half below the surface on the bottom of the sea. No sunlight ever reaches so far down. No plants grow in the absolute blackness. Only a little food trickles down—mostly bits of dead or decaying surface life and some animal droppings. And the temperature is just a few degrees above freezing. No wonder everybody thought that very few animals could live in such a dark, cold world. But that was before 1977, when a group of scientists aboard a research ship visited a special area deep down in the Pacific Ocean.

This particular part of the ocean floor was cracked in places. Hot, molten lava had welled up through the cracks and cooled quickly into hard rock. But here and there hot springs had formed when water seeped down through cracks, was heated by the lava, and boiled up again.

The scientists were going to study these springs. What they found amazed them and the whole world.

To begin with, they lowered a camera attached to a sort of sled that could be guided by cables on board the ship. As the sled moved, its lights shone on the sea floor and the camera took pictures that were made into slides.

The scientists watching their first slide show had views of dark lava—and then they saw something no one had ever seen before. Huge white clams clustered around a hot spring. There were crowds of brown mussels, too, along with other creatures. Here lived a whole big community of animals where none were supposed to be.

Later, a small submarine called the *Alvin* took scientists down for a look. The hot water boiled up through holes, called vents, in the rock. Around one vent something that looked like a dandelion was floating, attached to the bottom by a long, yellow strand. There were pink fish and white crabs and six-foot long red worms that lived in tubes they made of a substance that looks like white plastic.

Each community formed a kind of oasis in the ocean desert, and like any other group of animals, they needed food. But where did they get it? No one could even guess at first. Clearly they had plenty to eat, for they grew big and in large numbers.

On two more expeditions to the hot vents in the Pacific, scientists used *Alvin* to solve the mystery. The pilot inside operated a claw at the end of an arm outside. The claw scooped up specimens to be studied. Another tool was called a slurp gun—a sort of vacuum cleaner that sucked up small animals. One important instrument was a bacteria sampler.

Bacteria turned out to be the clue that solved the puzzle. Some of the bacteria the scientists found can take chemicals out of the hot water and turn them into food. The worms and clams and some other animals use this food and grow very fast. Other animals feed on the bacteria-eaters. Around some of the vents grow many of the small animals called forams. The bacteria nourish them, and they in turn are eaten by snails that bore holes into the foram shells.

The blood-red worms seemed hard to explain. They have no mouth or gut. Some are as much as twelve feet long, and so they must get plenty of nourishment. But how? Detective work showed that bacteria live in the worms' bodies. These microscopic chemical factories turn out food for the worms. But they don't do it for nothing. They must get something in exchange. What could it be? The answer comes from the color of the worms and their blood. It is red, like our blood, because it contains the same red substance that carries oxygen around in our bodies. We get oxygen from the air we breathe. The worm's body has a sort of plume at the tip which absorbs oxygen from the water. Then its blood carries the oxygen to the bacteria which must use it in producing food for the worm. The bacteria and the worm nourish each other, and that is why these amazing creatures have no need for a mouth or gut.

Inside its foot-long white shell, a clam's body is also bright red. It, too, depends on bacteria for its food and helps them out with oxygen.

More than two hundred kinds of bacteria live in and around the vents. You can't see a single one without a microscope, but there are three or four million in a teaspoonful of water. Billions of them form clumps on the rocks where fish feed on them. There is a shrimplike creature who has a special tool for collecting bacteria. Growing from its head are stalks tipped with combs which scrape its food, the bacteria, off the rocks.

The boiling water that pours from the vents quickly cools in the cold, surrounding sea. Where it is still warm, close to the vents, most of the animals make their homes. There, white crabs hunt for food. But they can somehow plunge into the hot water and come out unharmed. One eel-like vent fish also can stand a lot of heat. It was so strange and unusual that the scientists could not tell what its nearest relatives were.

Most of the vent animals do have relatives that live in the upper regions of the sea where all life depends on the energy of the sun. How did these deep-water cousins come to exist where life depends on the earth's inside heat? How do they spread from an old vent to a new one? Old vents disappear after about ten years. Perhaps by the time you read this page scientists will know some of the answers.

The Case of the Walking Flower

Is that a garden you see on the ocean floor? Tiny bushes seem to be growing near a large flower with long, fat petals. Suddenly the flower begins to turn upside down. Standing on the tips of its petals, it slowly walks away.

The flower that walks is a *sea anemone*. Although it looks like a garden anemone, it is not a plant. It is an animal. Around its mouth, looking like petals, grow tentacles that capture its food. Usually the sea anemone sits firmly attached to a rock. To walk, it lets the muscles in its base relax. The base then expands like a half-filled balloon and helps it to move lightly, as though it were stepping along on delicate legs.

Sea anemones come in dozens of wonderful colors and shapes and sizes. Some are as small as a dime and others stretch out three feet across. Some use their thick bases for creeping, the way a snail creeps. Others turn somersaults, bouncing on their tentacles from one place to another.

One kind of anemone lives in a group with others. If two members of the group feel crowded, they may compete for the same space. They push each other with special white-tipped tentacles that they use only in fights, not to catch food.

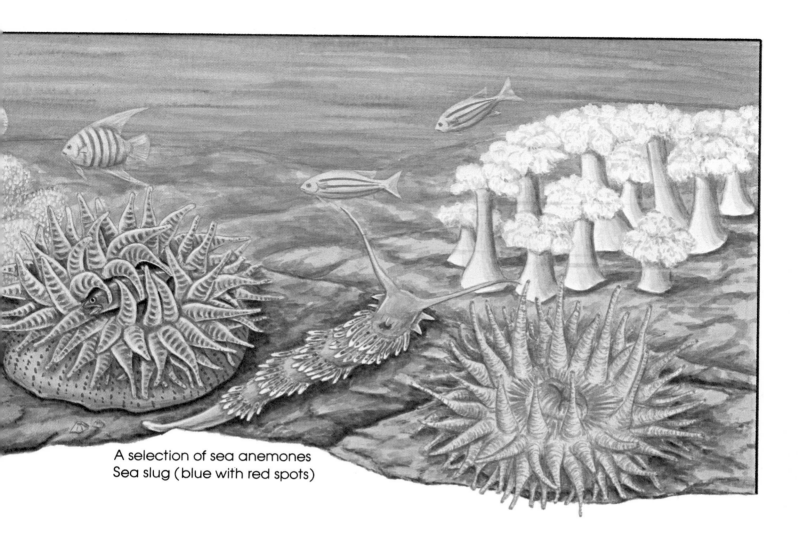

A selection of sea anemones
Sea slug (blue with red spots)

All sea anemones have enormous appetites. When a fish or shrimp swims too close and brushes against a tentacle, little poison darts shoot out. These paralyze the prey, which the tentacles then stuff into the anemone's mouth.

The poison darts are like tiny needles at the ends of hollow threads. The threads lie coiled up, like springs, inside capsules in the skin of tentacles. Each capsule has a lid with some sort of trigger. When a fish touches the trigger, the lid flies open. A coiled thread shoots out. The dart pierces the fish's skin and injects the poison.

Scientists still don't know all the secrets of the poison dart capsules. Why don't they sting a creature called a sea slug when it eats an anemone? Does the slug have some sort of anesthetic that keeps the capsules from opening? Even more remarkable, the capsules are not digested in the slug's stomach. Instead, they travel to the tips of fingerlike bits of flesh that stick out of the slug's back. Now if the slug is attacked, the second-hand poison darts shoot out and paralyze the attacker.

In some places the whole sea floor looks like a garden, with anemones and what appear to be delicate ferns and small weeds swaying in the water. The ferns and weeds are really animals. But they aren't single animals. Scientists say they are colonies of tiny creatures that live and grow together.

If you look at one of these living weeds under a magnifying glass, you can see little bumps on its branches. Each one of the bumps is a *polyp*—a single animal with connections to all the others. Together they form a group that looks like the buds and stems and branches of a plant.

In the water, the tip of the bud opens, and out pop fine, waving threads like the petals of a flower. The open tip is the polyp's mouth. The threads are tentacles that spread out to capture food.

The polyps in these colonies have many relatives. One kind forms a pale growth that looks like a ghostly hand sticking up out of the ocean floor. Fishermen call it Dead Man's Fingers.

Other polyps create stony skeletons around their bodies. They are the coral animals. After a coral polyp has attached itself to something firm, its base begins to take in a mineral which is dissolved in the sea water. The mineral, calcium carbonate, hardens and forms a cup around the bottom part of the polyp. The top part then pokes a bit of its body out over the rim of the cup and joins the body of another polyp next to it. One after another the coral animals build hard skeletons and join bodies. Now when food is digested in one polyp's gut, it can pass along and be shared by others in the colony.

Coral animals produce eggs which are fertilized and released into the sea. There they float about till tiny new polyps develop and settle down on the sea floor to form the beginnings of new coral colonies.

When a coral polyp dies, another is likely to build its stony cup on top of the first one's skeleton. So a colony can keep on growing until the corals form whole stony walls called reefs just below the ocean's surface. One scientist wondered how many corals had to work at building a reef seven feet long, five feet wide and one foot high. The answer: More than a million.

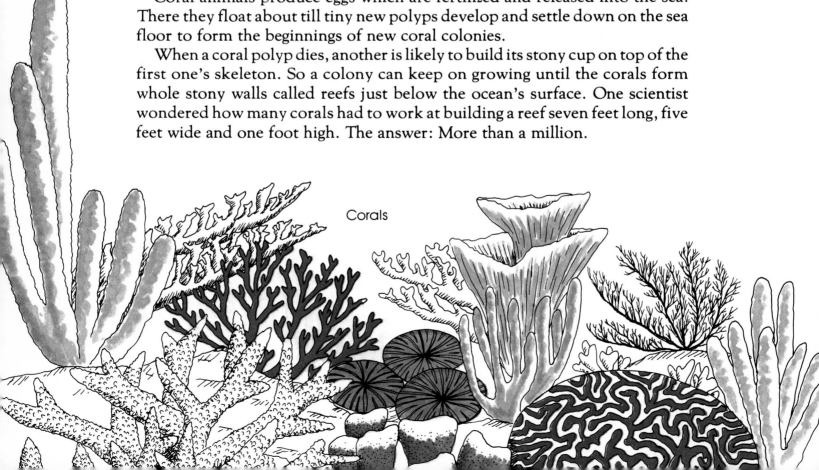

Corals

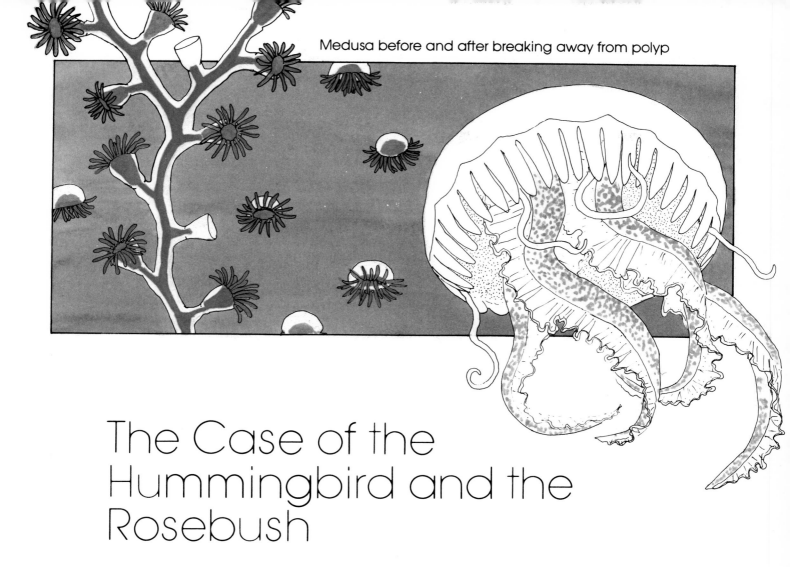

Medusa before and after breaking away from polyp

The Case of the Hummingbird and the Rosebush

Long ago, the Greeks believed there was a dangerous creature they called Medusa. She had a human face and body, but snakes grew on her head instead of hair. Medusa was what scientists thought of when they were choosing a name for a sea animal that swimmers don't like. The medusa of the sea is shaped like a small dome or bell, and all around its rim hang snaky tentacles. The trouble with a medusa is that these tentacles carry stinging darts. The darts are used for paralyzing fish or other prey that the medusa eats. But they also give an irritating injection to swimmers who happen to touch them.

Another name for a medusa is jellyfish. Its body is indeed made of a substance like jelly between two layers of transparent skin. Not all jellyfish are alike. Some produce eggs that develop into new little ones. Others do not. Instead, they sprout little bumps that look like buds. These buds grow and turn into young jellyfish. There are still other kinds of tiny medusa that fooled scientists for a while. They produced eggs, and the eggs did not seem to develop into young jellyfish. Why not?

23

The answer came when people studied the little animals called polyps that grow in colonies and look like ferns and other plants on the sea floor. A colony gets bigger as new polyps bud off from older ones. Then, every so often, a very strange thing happens. A different and special kind of bud forms. After a while it pulls itself loose from the plantlike stem. But it does not sit still on its base, with tentacles waving above it like its sister polyps. Instead, it floats off by itself, with tentacles hanging downward. This upside-down polyp is a medusa.

Later, the medusa produces eggs which settle to the sea floor. But a medusa egg of this kind does not develop into a new jellyfish. Instead, it turns into a rightside-up polyp. Each polyp grows, buds, and forms a new colony. Sooner or later this colony produces a medusa which produces eggs which turn into polyps.

This process, says one scientist, is as strange as if a rose bush gave birth to a hummingbird and then the hummingbird's eggs hatched into rosebushes.

The tiny stinging jellyfish have relatives that can give swimmers a really dangerous injection of poison. One, called the sea wasp, is shaped like a box, not a bell. The poison from its darts can kill a person in a minute or two.

The large, beautiful Portuguese man-of-war is sometimes called a jellyfish. Actually it is a whole colony of medusas and polyps that live and float on the sea together. All of them cooperate in catching food, digesting it and sharing it. The man-of-war gets its name because the part that shows above the water is shaped something like a sail of an ancient war ship. This bright blue sail is filled with gas, which keeps the colony afloat. Attached to the underside of the float are dozens of tentacles. Some of them trail out as much as one hundred feet in the water, searching for food. They have stingers that paralyze prey—and also give poisonous and painful injections to any swimmer who tangles with them. Other tentacles carry a fish or other prey up to the polyps attached to the float. A whole group of the polyps fasten onto the fish. Juices from their bodies digest it. Soon there is nothing left but bones and some skin, and the food circulates to all the members of the colony.

Like some other sea creatures, the man-of-war often allows certain fish to swim along with it unharmed. Just why they aren't stung and eaten is a puzzle.

When the little submarine *Alvin* took scientists more than a mile down into the deep sea, they saw what they called the Dandelion Patch. Like yellow-headed flowers, creatures covered with stubby tentacles bounced and swayed at the ends of threads attached to the rocks. What were they and what kept them afloat? The scientists discovered that the "dandelion" is a relative of the

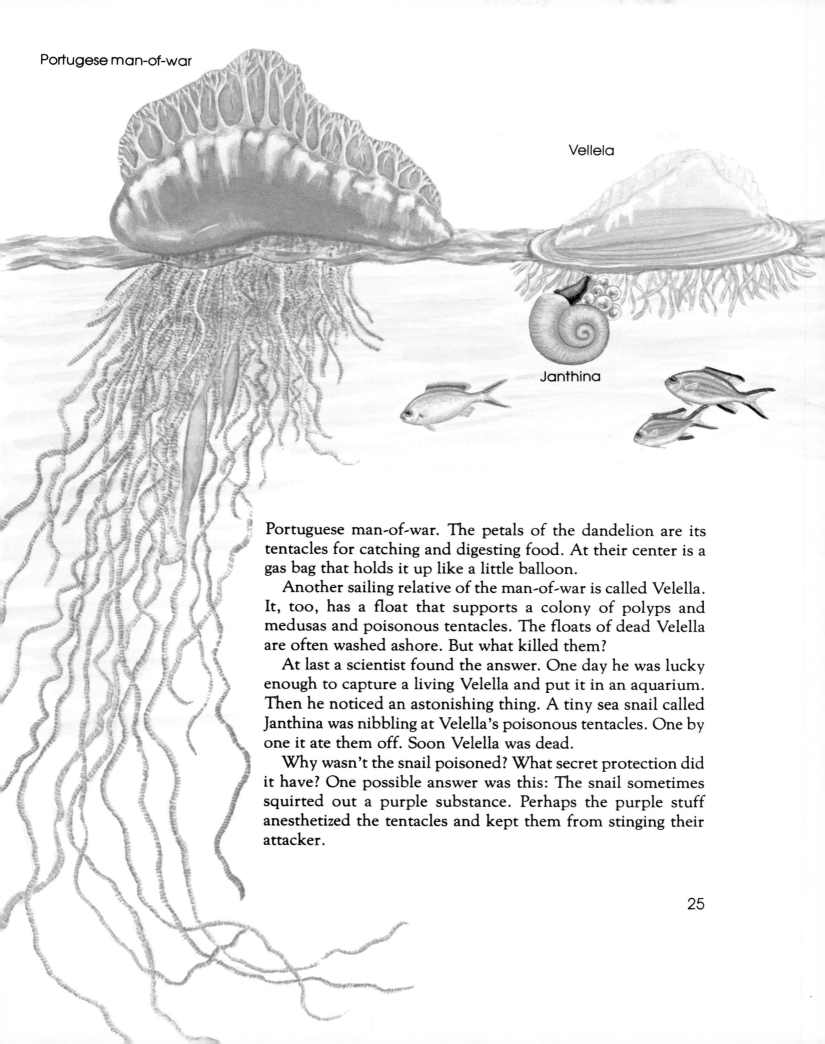

Portuguese man-of-war. The petals of the dandelion are its tentacles for catching and digesting food. At their center is a gas bag that holds it up like a little balloon.

Another sailing relative of the man-of-war is called Velella. It, too, has a float that supports a colony of polyps and medusas and poisonous tentacles. The floats of dead Velella are often washed ashore. But what killed them?

At last a scientist found the answer. One day he was lucky enough to capture a living Velella and put it in an aquarium. Then he noticed an astonishing thing. A tiny sea snail called Janthina was nibbling at Velella's poisonous tentacles. One by one it ate them off. Soon Velella was dead.

Why wasn't the snail poisoned? What secret protection did it have? One possible answer was this: The snail sometimes squirted out a purple substance. Perhaps the purple stuff anesthetized the tentacles and kept them from stinging their attacker.

Clams digging into sand

Some Strange Equipment

One minute it is there; the next it is gone. How can a clam disappear from sight so quickly?

Clams and some of their relatives dig themselves down into the sand on the sea floor. There they can stay for a long time, unless some other animal digs them out. They can live this kind of life because of two wonderful body parts.

First, an organ called a foot pokes itself out from between the clam's two shells. This foot wiggles its way downward, stretching out until a bit of something like a leg is showing. The foot swells out to make a kind of anchor in the sand, and the muscles in the leg contract. But they do not pull the foot back into the shell. Instead, they pull the body of the clam toward the foot.

Digging and pulling, the foot hauls the clam down, step by quick step. Meantime, another piece of equipment has appeared at the opposite edge of the two shells. This one is made of two hollow tubes, called siphons, joined together. The farther the clam goes into the sand, the longer the double siphon stretches out. The open tips of these tubes always stay just above the sand, so that water can enter one of them and come out through the other.

It may seem that the clam has some sort of pump inside to keep the water circulating. Instead, the work is done by multitudes of microscopic, hairlike parts called *cilia* that line the siphons. The cilia rhythmically beat back and forth to create a current in the water. The current sweeps oxygen and particles of food down into the clam's body. Other cilia keep the water moving up the other siphon to carry away wastes.

Many sea animals have cilia that channel food to their mouths. Some of the tiny hairs can even push away anything that isn't just right to eat. One small creature called a *ciliate* lives inside a glassy house it builds around itself. The cilia at the top of its body beat, one after the other, round and round. Like a helicopter rotor, they make it rise and hover in the water. They can also reverse and make it sink.

A barnacle has a hard outside covering, and you may think it is a shellfish when you first see it. In fact, it is a relative of the shrimps. Young barnacles develop from eggs inside their parents' bodies. For a while after they hatch, they float about in the sea. Then they settle down, head first, on something solid—a rock or a boat or even a whale. Immediately, each young one squirts out a special cement that holds it tight to its perch. Next, with a substance it takes from the water, its body creates a protective covering. Six of its feathery legs now wave about in the water while the rest of it stays upside down in its solid little house. The legs open and close like a trap, two or more times a second, catching particles of food which cilia on the legs shove down to the mouth.

The fiddler crab, also a relative of the shrimps, feeds on scraps that it finds in mud on the sea floor. With spoon-shaped hairs on little organs near its jaws, it separates the food from the mud in its mouth. Then it shapes the mud into a ball and throws it away.

Fiddler crab   Barnacles feeding

Some of the 1800 kinds of starfish have five arms equipped with stumpy little tube-shaped feet that end in suckers. These tubefeet dig into the sand as the starfish hunts for food. When it finds a clam, the suckers on one arm press against the top shell. Those on another arm attach to the bottom shell. Then they begin to pull, trying to make the clam open up. But the clam has a strong muscle that holds its shell tightly closed. The tug-of-war goes on, sometimes for hours. Finally the clam gets too tired, and the starfish wins. Then, when the clam's shell falls open, the starfish does a remarkable thing. It turns its stomach inside out on top of the clam, and the stomach juices digest the meat.

Sometimes a starfish tackles a different creature—one that has a very strong muscle. No matter how hard the starfish pulls, it cannot make the shells open. So it has to give up and walk away on its stumpy tubefeet. The feet do not seem to move in rhythm. Instead, they march "like an army out of step," said one scientist. The suckers on the feet produce a sticky substance that helps the starfish to climb up underwater rocks. Some feet cling to the rock, while others step upward. This is slow work. A starfish can only climb about six feet in a day.

One type of starfish swims instead of walking. It paddles with four of its five arms and keeps the fifth pointed straight ahead. Its cousin, the feather star, has ten arms that flutter and move it through the water.

Starfishes and a yellow featherstar

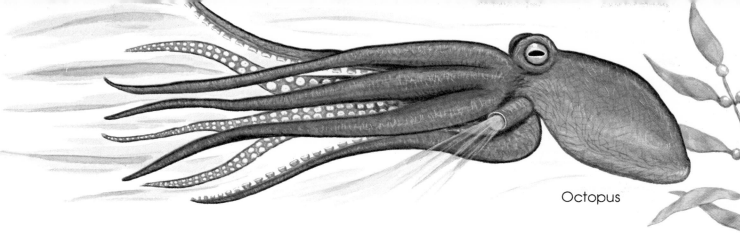

Octopus

An octopus is related to a clam but not to a starfish. It has a siphon and its eight arms are equipped with suckers which catch and hold its prey. Its siphon works like a jet engine. When the octopus points the siphon straight out in front of its body, and squeezes out a burst of water, the water jet sends the octopus sailing backward. By pointing the siphon in different directions, the octopus can move sideways or forward, too.

A long, skinny creature called a sea lamprey has a round sucker, which is also its mouth, at the very front of its head. Inside the mouth are rows and rows of sharp teeth and a tongue that works like a file. After a lamprey attaches itself to a fish, teeth and tongue go to work, rasping away skin and flesh. Then, still attached, the lamprey sucks out blood. Holding tight with its teeth, a lamprey may be hauled along by a large fish for several days until the fish dies from lack of blood.

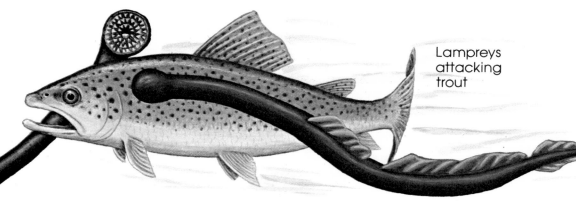

Lampreys attacking trout

One type of sea snail has special equipment for eating oysters. First, it opens its mouth and uses an organ called a *radula*. The radula is covered with sharp, five-pointed teeth that it rasps back and forth against the oyster's shell, boring a hole. If a tooth breaks off, the snail swallows it. As it scrapes, the snail pours out a chemical that dissolves the oyster's shell. At last, there is a hole all the way through. Now the snail pokes its snout into the borehole and eats, using side teeth in its mouth to cut up the meat which is then sucked into its stomach.

Sometimes a crab discovers the snail eating and comes to share the meal. If it doesn't get a tidbit, it may bite off the snail's snout. But the snail grows a new one.

Perhaps it may seem by now that the sea is a crazy world with one kind of animal constantly killing others, always in danger itself. It is true that only a few creatures are free to roam unharmed—except when people hunt them. But life everywhere in the sea fits together, so that thousands of animals can live in communities, and one kind never slaughters another just for the sake of killing. Even when a great white shark gets excited and wildly bites its prey in two, nothing is wasted. The uneaten parts end up as food for smaller animals.

A kind of rhythm in the animal communities keeps them steady. For example, when certain starfish mate and produce eggs, they stop eating. This gives the animals they feed on a chance to grow. Codfish, which live near the sea bottom, lay millions of eggs. So do herring, which live higher up in the water. The codfish eggs rise and float near the surface. There the herring gobble up quantities of them, but they don't succeed in eating them all. The herring eggs, millions of them, sink to the bottom where the cod eat a lot of them, but not all. After the eggs hatch, however, the young fish trade places. The young cod go to the bottom where they are safer, and the young herring swim up out of reach of the cod. And so the rhythm of life in the sea preserves both kinds of fish.

You can tell what a fish eats by looking in its mouth. The ones that feed on seaweed have sharp teeth that are good for cutting. Some also have grinders in their throats to chop up tough plant food. Fish that eat other fish have pointed teeth arranged so that they can grasp and tear their prey easily. The fish that eat clams or other creatures with shells have heavy teeth for crushing the shell. One called the wolf eel, though its is not an eel, feeds on spiny sea-urchins. With powerful jaws and heavy grinding teeth the wolf eel crunches the sharp spines that protect the urchin's soft body. Why don't the spines puncture the fish's mouth as it eats? Sometimes one does poke through its lip. But the teeth cover so much of its mouth that there is little space for bare flesh. The sea urchin itself has powerful teeth for eating tubeworms and barnacles and other tough creatures. Another kind of urchin feeds on sea plants rather than other animals. It can chop through solid stalks by sliding its five teeth up and down in special holes in its jaw.

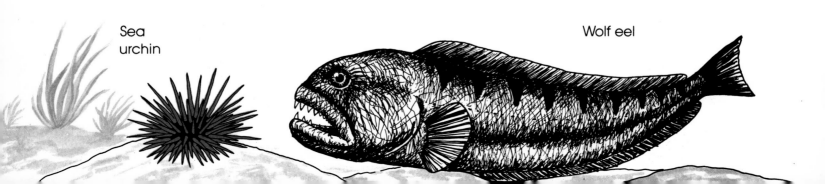

Sea urchin

Wolf eel

Southern right whale showing baleen

Some fish with small mouths have a kind of loose, bony curtain in front of the gills. The curtains are called *gill rakers*. These fish eat small plants and animals of the plankton. When the fish takes a mouthful of food and water, the gill rakers act like a sieve. They trap the food, while the water is squeezed out through the gills.

Some whales have similar curtains in their mouths. Although the curtains used to be called whalebone, they are really strips of a substance called *baleen*, which is the same stuff as your fingernails. The strips, or baleen plates, hang down from the sides of the whale's upper jaw, and each strip is fringed. This whole arrangement looks rather like a moustache. So a scientific name, meaning moustache-whale, was given to the kind that have baleen plates.

The fringed strips in the whale's enormous mouth filter out the small plankton animals that it eats. But a whale is not a fish, and it has no gills. So how does it get rid of water that comes in with its food? Does it have to swallow huge gulps and then belch the water out?

Scientists supposed that a baleen whale could somehow squeeze the water through its lips while the fringes trapped and held the little plankton creatures. Then its tongue would lick the food off so it could be swallowed. Finally a man who was studying humpback whales in the Atlantic Ocean made a surprising discovery. He knew that there is a large bulge under the humpback's throat. The skin over the bulge seems to be able to expand like an accordion. But how did that help the whale to take in an enormous lot of water at one time? The scientist found a clue when he was examining a dead humpback whose head was tilted back, mouth open. The whale's tongue, too, was tipped over backward—and it was hollow!

This seemed to be the answer. The hollow tongue could hold a huge quantity of water and food, and tipped back it could make the bulge under the throat expand. Then, with a tremendous squeeze of muscles, water was forced out of the sides of the mouth, while the food stuck to the fringes of the baleen plates. Now it could be licked off and swallowed.

Some whales do not have baleen plates. They eat fish and squid and other kinds of animal, large and small. The killer whale is the hungriest and the fastest of them all. It eats almost anything it finds, and it probably got its name because its speed and its huge appetite made it seem vicious. The truth is it has a very good temper when it lives along with captive dolphins that put on shows for people. Its trainers keep it well fed, and it learns tricks easily. Human swimmers can even ride on its back.

Killer whale chasing sealion

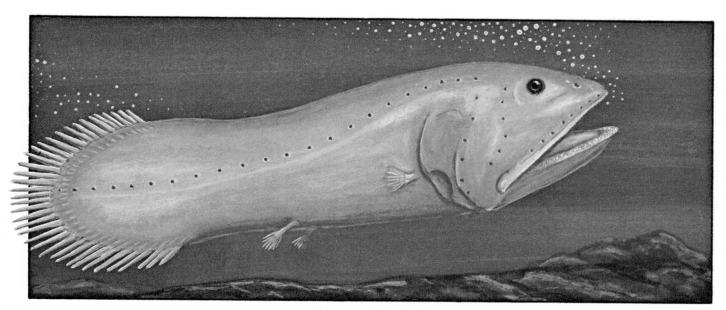

Whalefish showing portion of lateral line

Senses and Extra Senses

If a whale had ears like a mule or a man, they would spoil the shape of its head, which is streamlined for fast, easy swimming. Its outer ears are just small holes that you can hardly see, a little way behind its eyes. Even with such a small opening to catch sound waves, the whale's sense of hearing is very sharp. Humpback whales have a curious hearing aid in their ears—a long plug of wax that conducts sound waves to their ear drums.

Fish do not have ear openings to the outside of their bodies. Some of them get sound signals from their swim bladders. Because a fish is mostly water, sound waves go right through it. But the swim bladder is filled with gas, and sound waves make it vibrate. These vibrations trigger nerve signals to the fish's ears.

Many fish detect sounds with what is called the lateral line. This a row of tiny holes in the skin along its side. The holes open into a tube that is lined with little tufts of hairlike material connected to nerves. Sound waves make these hairs vibrate, and the nerves carry the message to the brain.

Herring and some other fish have chalky ear stones, but not for hearing. The stones roll around in the ear chamber, touching sensitive hairs. Nerves connected to the hairs notify the brain that the fish is tilting up or down or to one side or the other. This gives it a sense of balance and direction.

The lateral line also may help with balance. Some people think it may even help a fish to know how fast it is traveling. Perhaps its nerves pass along other sensations, too. One scientist says that if we had a lateral line, and if our eyes were covered and our ears plugged, "we could sense exactly the position of a bee buzzing around our head."

Other sea creatures have balance organs in curious places. The opossum shrimp gets balance from an organ in its tail. The sea gooseberry's organ is on top of its round little body. The octopus and its relatives have balance organs in their heads, and their arms and tentacles help to tell them whether they are right side up. The arms give them a delicate sense of touch, too. Their eyes, very much like those of land animals, are large and sharp-sighted.

Shrimps and their relatives have eyes more like those of a housefly. Each eye is made up of many separate seeing parts. And so they can look in all directions without turning their heads.

A scallop has glowing blue eyes just inside its shell—as many as a hundred of them—placed so that it can sense everything going on in a half-circle in front of it. The scallop's eyes do not see as well as those of a shrimp, but they can detect motion. At any sign of danger the scallop snaps the two halves of its shell shut, then open, then shut—in little bursts. This forces water in and out, so that the scallop goes bouncing away through the water.

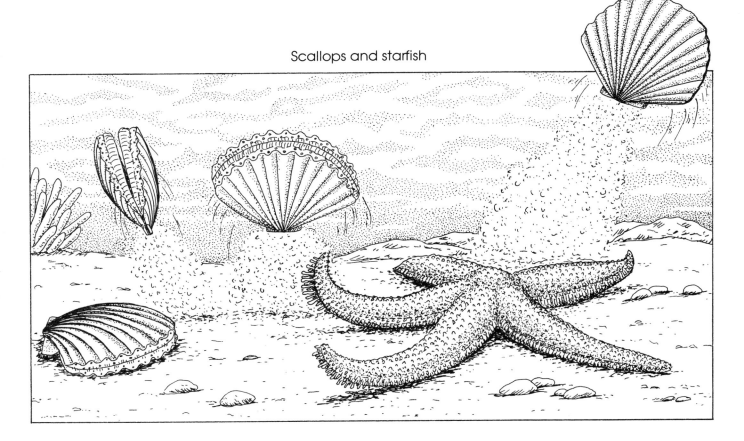

Scallops and starfish

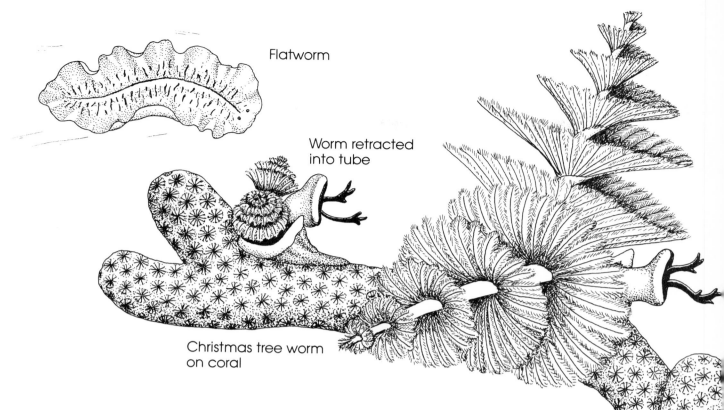

Flatworm

Worm retracted into tube

Christmas tree worm on coral

Some sea creatures do not have fully developed eyes. Instead, they have sensitive eye spots. A worm that looks like a Christmas tree is one of these. It lives inside a white tube that it builds for itself. When it comes out to eat and breathe, it unfolds a number of feathery stalks that collect food and oxygen. Eye spots on the stalks detect the shadow of any enemy, and the worm instantly draws back into its tube.

Curious little animals called flatworms have a few eye spots at their front ends, just above their mouths. They paddle themselves along with hairlike cilia on the underside of their bodies and with the help of the eye spots, they can do a surprising thing. They can learn to find food in a maze. In a short time they will remember to paddle along white lanes in the maze and stay out of dark ones. That is surprising enough. Even more astonishing is the way their memory works. If one flatworm is cut up and fed to another, the second one gets the first one's memory. It will follow the correct lanes in the maze—even though it has never been in a maze before.

The sea snail that eats the Velella jellyfish has no eyes. How it locates its prey is a mystery. Perhaps it has spots on its antennas that are sensitive to light and shadow.

Herring have very sharp eyes. They and about 4000 other kinds of fish travel together in groups called schools. Their eyes are specially set in their heads so that they can see their companions on both sides and also in front. This helps to keep the school together and also helps to give warning that an enemy is approaching.

If a shark appears on one side, the whole school immediately changes direction. How do individual fish sense what all the rest are going to do? Researchers don't agree on the answer, and no one is really sure. Possibly the lateral line helps. The hairs in it are sensitive to currents and vibrations. So when one fish changes direction, the slightly changed current warns the next and the next and the next, and all quickly get the signal.

Schooling helps fish to find mates. It also confuses enemies who can't decide which in the flashing group to grab and eat. So the whole school may escape. A big enemy, of course, can just open its mouth and swallow a lot of fish at once. But there are many more small fish than the big meat-eaters need. Schools of herring are often so huge that they form a solid mass stretching out for miles. In the midst of such a school fishermen in a rowboat cannot use their oars.

Some big meat-eaters have a keen sense of smell. They can detect a school some distance away. The hammer-head shark has a special advantage. Its head really is shaped like a sledge hammer, and it has sense organs at both tips of the hammer. That means it can easily smell prey on either side of it, not just in front.

Hammerhead shark and school of snappers

Salmon

A shark is a fish with a special kind of skeleton made of stuff called cartilage instead of bone. (Your nose is made of cartilage.) Many fish with bony skeletons also rely on their sense of smell. They detect smells with sensitive spots in two nose pits on their heads. Water flows into the front pit and out of the one behind it. The sensitive spots can respond to particles of stuff in the water. This delicate sense of smell in salmon amazed scientists who tested it. They found that a salmon can detect the scent of the water in the stream in which it hatched from an egg. Not only that, this fish has a long memory for smell. It leaves its home stream when it is tiny and swims out into the ocean. There it stays until it is old enough to mate. Then it returns to the very stream it left years before.

Possibly a salmon tastes as well as smells its home stream water. Many fish have a delicate taste sense which helps them to hunt. A fish called the whiting has taste buds on its fins. The codfish has taste buds on its fins, its lips, and its body, including a kind of whisker that hangs down from its chin. One deepwater fish has fins with very long, thin parts called rays. It uses the rays both to walk in the mud on the sea floor and to taste the mud, hunting for food.

Deep-sea ray fish

There are at least two sea animal senses that human beings lack. One is an electric sense that helps sharks to locate the fish they eat. If a fish just wiggles its fins, it creates a very tiny electric current in the water. The shark's radarlike sense can pick up the signal and it dashes after its prey. Exactly how this radar works is something scientists don't yet know.

Researchers do know that the yellowfin tuna has a special magnetic sense. Tests have proved that tuna are sensitive to magnets, and a magnetic mineral called magnetite has been found in their brains.

Yellowfin tuna travel long distances across the Pacific Ocean from one feeding ground to another. There seems to be no doubt that their magnetic sense acts like a compass to guide them day and night without sight of land or sun or stars.

The bluefin tuna also travel thousands of miles in the Atlantic Ocean, north in summer and south in winter. By the time a bluefin is fifteen years old, it has probably gone a million miles, swimming day and night. Bluefin cannot stop to rest because they need more oxygen than most other fish, and they are built differently. They have no way of pumping water over their gills. Instead, they swim constantly with mouths open to keep water flowing and bringing in the oxygen they must have. The routes the bluefin take in their travels seem to follow warm currents in the ocean. They may also rely for help on a magnetic sense like their relatives the yellowfin.

Particles of magnetic material have also been found in dolphins' brains. The particles are surrounded by nerves, and this seems to mean that they have a magnetic sense of some sort. Possibly salmon, too, have a magnetic sense that guides them thousands of miles through the ocean to the coasts where they then sniff out their own home streams. Some scientists believe salmon may also navigate by observing the sun and stars.

Green turtles live most of their lives in the sea. They visit the shore only to mate and lay eggs in nests that they dig in the sand. When the baby turtles hatch, they hurry across the beach and swim away. After several years of traveling about in the water, the young turtles are grown and old enough to mate. At that time they always come back to the exact same beach where they were born. Scientists hope someday to find out what this turtle homing sense is and how it works.

Are other unknown senses hiding in some of the sea's smallest creatures—the foraminifera? A foram has no eyes or nerves or brain. But some forams can build houses for themselves from bits of glassy stuff that comes from broken sponges. One will shape the little sticks into a tent. Others build shelters like nests with criss-cross woven sides. What is even more amazing is that they arrange the glassy fragments according to size. Without any of the ordinary senses, how does a foram perform this precise work? That is a secret still waiting to be unraveled.

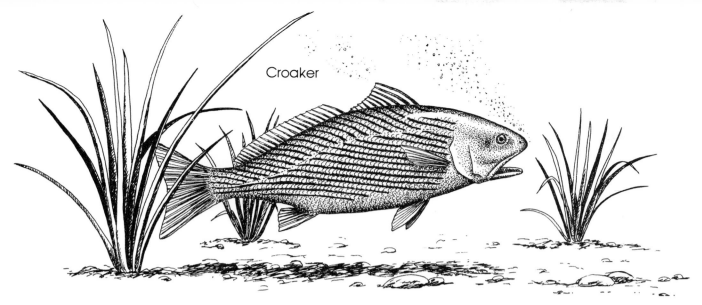
Croaker

Things That Go Beep in the Deep

Most people used to think the world beneath the surface of the ocean must be perfectly quiet. Then came the Second World War. The war brought a great deal of destruction, but it had a constructive result that was a surprise. Scientists who study the sea discovered at that time some things they had not known before.

During the war submarines began to prowl underwater, hunting for ships to sink. No one above the surface could see or hear them. In order to locate enemy submarines, the United States Navy used an invention that could pick up sound waves traveling through water. These sound detectors did locate submarines. They also detected mysterious grunts, chirps, pops, hoots, whistles, beeps, and loud noises like gunshots and jackhammers. Something down there was not being quiet. But what?

Later, scuba divers discovered that a fish called the croaker made jackhammer sounds that actually hurt their ears. Scientists finally figured out how a croaker does it. Like most fish, this one has a kind of balloon inside it called the swim bladder which is filled with a special gas. Like a balloon, the swim bladder helps to keep the fish from sinking when it isn't actively swimming. The swim bladder also acts as a sort of drum. The croaker can expand and contract strong muscles attached to the side of the swim bladder. The muscles move back and forth very fast—24 times each second—making the swim bladder vibrate. These vibrations travel out into the water and quickly reach the diver's ears. Sound travels five times as fast in water as in air.

Vibrations also caused the other sounds that were so puzzling. Spiny lobsters make squeaky chirps by rubbing their antennas against pads under their eyes. The pads are made of row on row of horny material shaped like tiny saw blades. Why the lobster makes the saw teeth vibrate is not clear.

Delicate crunching sounds come from parrot fish nibbling away the hard outside coats of corals to reach and eat the coral animals inside. One parrot fish can grind up as much as thirty pounds of coral in a year, dropping it to the sea bottom where it forms coral sand.

Little damsel fish go *click-click* when they are annoyed. They make the sound by rubbing together the teeth that line their throats. Some fish rub the stiff rays of their fins together—*br-r-r-r*.

The multi-colored parrot fish noisily crunches on coral while the clicking damsel fish swim away; and at the sea bottom sits the spiny lobster making its squeak-like chirps.

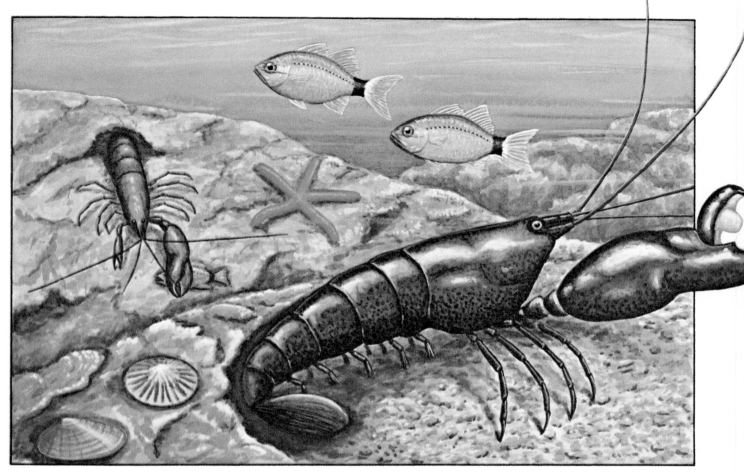

Pistol shrimp

The gunshot sounds come from the pistol shrimp. Two of its legs have pincers on them, and one pair of the pincers is shaped in a remarkable way. In one claw of the pincer is a small pit. The other claw—a sort of finger—has a knob that just fits into the pit. The pistol finger can snap the knob down with amazing force. If you are holding one of the shrimps when it shoots, your hand feels a sharp pain like a sting. Any scientist who studies this little creature has to be careful to keep it in a container that has no scratches or cracks. A shot from its pistol has been known to break a scratched bottle.

A pistol shrimp uses its claw to catch its food. If a fish approaches its burrow, the shrimp reaches out, touches the claw to the fish's head and stuns it by snapping the pistol finger. Large numbers of these shrimp sometimes live close together. The noise their pistols make can be heard even outside the water.

A hoot like a foghorn or a boatwhistle interested one group of scientists, and they decided to find out all about it. Who made it? Why? In time they found that the whistler was the toadfish. Its muscles could drum against the swim bladder as many as one hundred times a second.

A male toadfish makes a nest in sand or in an empty shell—even in a tin can that someone has tossed from a boat. For their experiment the scientists provided the fish with a can. Then they listened.

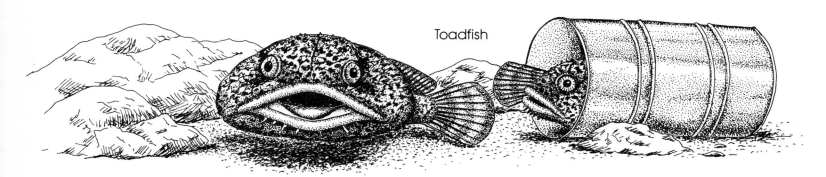
Toadfish

The male toadfish, they discovered, makes two kinds of sound. It gives a whole lot of grunts that add up to a growl when it warns other males away from its nest. To attract a female it makes its boatwhistle calls.

After the female places eggs in the nest, the male stands guard and then stays to protect the young fish for a while after they hatch. Sometimes a nesting father seems to be taking care of young ones of different ages. Did the eggs hatch at different times? Finally the observers found out that several female toadfish may answer the boatwhistle and place eggs in the nest at various times.

The toadfish's loud sounds—only the pistol shrimp makes louder ones—are as useful as a foghorn is to a ship. These fish live in water that is often murky and hard to see through. The hoots and growls tell mates and enemies where the toadfish male can be found or avoided.

Many other fish make sounds, but most of their messages are still a mystery. The places where large communities of different sea creatures live must be as noisy as a city street. No one knows whether they can all hear each other's sounds. Probably not. But each creature, when it is young, must learn which sounds are important for its kind—just as you respond when a friend calls your name, even though noisy cars are passing by.

The clicks, whistles, squawks and barks that bottlenose dolphins make sometimes sound almost human. Scientists who listened to them began to wonder if dolphins could learn to talk. After all, they are mammals, not fish. They have lungs and must come to the surface of the water to get oxygen from the outside air, and mothers nourish their babies with their own milk. Captive dolphins cooperated with trainers and played with them and quickly learned tricks. They even learned to make new sounds, almost as if they were imitating what their land-mammal friends were saying. But so far no dolphin has used human words.

Long ago, people noticed that these clever animals can find food in dark or muddy water, and that they can swim at night without bumping into anything. Scientists have now discovered that a blindfolded dolphin can do the same thing. Does it have some kind of special sense that guides it?

The answer is that it uses echoes to tell it where things are in its path. As it swims, the dolphins makes sounds which travel through the water. When the sound waves reach a fish or an obstacle, they bounce back. These echoes strike the dolphin's ears and tell it that food or even a fine net lies ahead.

A dolphin doesn't really have a voice. Unlike human beings, it has no vocal cords in its throat. But something in its body must cause the vibrations that create sound waves. The best guess is that the sounds are made in air spaces in the top of its head. Above these spaces is an opening called the blowhole through which the dolphin breathes. A burst of air blown out of the blowhole creates the sounds. How the dolphin can change from clicking to squeaking to whistling is still a puzzle.

Bottlenose dolphins

Humpbacks travel together in groups.

Dolphins are members of a large group of mammals—the whales. The big whales, like their small dolphin cousins, make sounds that scientists believe are signals to each other. The meanings of all their clicks, groans, and squeaks haven't been figured out, but we do know that the sounds travel great distances through the water. Whales that are miles and miles apart can hear each other. No one knows what difference this may make in their lives.

Some whales sound almost like musical instruments. Most musical of all is the humpback. Its songs have rhythm and melody that can be written in notes just as if they were to be played on one of our instruments.

Humpbacks are very large baleen whales with two long, thin flippers. When they are studied, they get used to people and even seem to enjoy being watched. If a researcher bangs on the side of a boat, a humpback—or perhaps several—may swim up close. Their big heads suddenly pop up near the boat, dark and warty-looking, like dill pickles. Each of the bumps on their snouts holds a stiff hair. All the rest of a whale's body is hairless, which makes swimming easier. Why do these few bristles remain? Possibly, these hairs act like antennas, and give the whale information.

Humpbacks travel together in groups called *pods*. So do other kinds of whale, but none of them sing such complicated songs as those of the humpback. All the whales in a particular humpback pod sing the same song, which has at least two, and sometimes as many as eight themes. In any one year, the order in which they sing the notes is always the same. But the next year the song may be somewhat different.

The humpback singers are all males, and they all perform in the same way when they are singing. Each one stands on his head in the water, about 150 feet down, and keeps his flippers stretched out. When his song is over, he dashes off to join the rest of the group. Everyone now agrees that he sings when he is looking for a mate. But it is not clear how he makes his music or why it has become so complicated and precise.

Male humpbacks singing

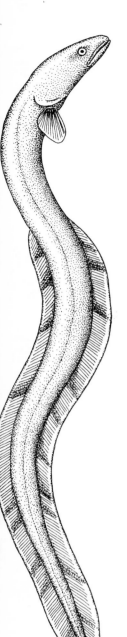

Baby eels in sargassum weed

Babies in Disguise

Eels have been one of the hardest scientific puzzles to solve. These long, thin fish look rather like snakes. People have been catching them in rivers and ponds on both sides of the Atlantic Ocean for hundreds of years. The fishermen always knew that big eels, when they got old, traveled down rivers and out to sea. But big ones never returned. The only eels that swam out of the sea and upriver were babies. Where did the babies come from?

Part of the puzzle did not fit into place until almost a hundred years ago. Meantime, scientists found something that they did not know was a clue. In an area of the Atlantic Ocean called the Sargasso Sea, far from land, there turned up a transparent fish about a quarter of an inch long. The little fish got a big scientific name—*Leptocephalus brevirostris*. But you don't have to remember those hard words, because the fish had its name for only a short while. In 1896, two people captured one of the little creatures and kept it in an aquarium. Gradually it grew and began to look different. Then one day they saw something no one had ever even suspected before. Their tiny fish had turned into a baby eel.

Baby eels are called elvers. The transparent early form is a *larva*, and more than one are *larvae*. How did eel larvae happen to be in the middle of the ocean? There was only one answer. Adult eels had to swim all the way down rivers, then more than a thousand miles to the Sargasso Sea. There they mated and left their eggs to grow into larvae. And after the larvae turned into elvers, baby eels would appear in rivers. But how did their parents find the way to the Sargasso Sea? And how did the elvers navigate back to rivers? That is still the eels' secret.

Many sea creatures start life as larvae that look very different from their parents. When scientists first began to study the plankton, some of the little drifters fooled them. One kind of crab's larva, seen under a microscope, looks like the weird monster at the right.

Crab

Crab larva

Only an experienced fisherman could guess what one kind of baby fish will look like when it is grown. It is only a tiny, transparent thing, about an eighth of an inch long, when it comes out of the egg. Surprisingly, the mothers of such small babies are ten-foot-long blue marlin that weigh about 900 pounds.

A mother octopus produces several thousand eggs which are attached to thin strands. Then she gathers them up and holds them in her eight arms until they hatch about a month later. All that time she eats nothing. When the babies struggle out of their egg cases, they look much more like their parents than many other larvae do.

One kind of squid—a relative of the octopus—lays hundreds of eggs packed together in one long, thin outer capsule that looks like a huge, transparent vitamin pill. Sometimes thousands of squids mate in one small area. The capsules, full of their eggs, form a glistening mat on the sea floor.

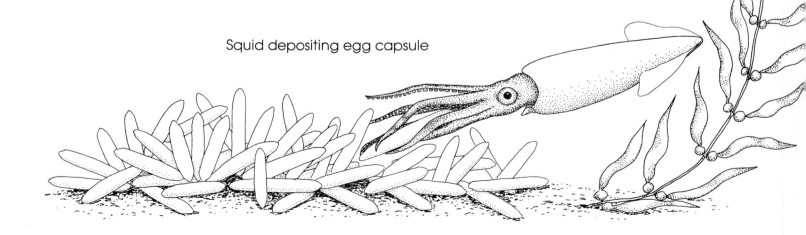
Squid depositing egg capsule

Young seahorses coming from father's pouch

Some animals keep their eggs inside their bodies, and the young are born alive. Barnacles hold the eggs inside their shells. One kind of arrow worm carries its eggs in a pouch. So does one kind of shrimp—the opossum shrimp. You can probably guess that it was named after land opossums whose young live for a while in their mother's pouch after they are born.

Father seahorses are the ones that have the pouch. After seahorses mate, the female places several hundred eggs in the pouch. When the eggs hatch, the young ones feed on a substance in the pouch, and the father swims around with a big bulge in front until the babies are able to take care of themselves. Then he wraps his tail around a bit of sea grass and squeezes the babies out. Immediately they dash to the surface of the water and gulp in a mouthful of air to fill their swim bladders. From then on they are as independent as their parents.

The female manta ray is a fish with a wide, flat body that ripples as she swims. Instead of laying eggs in the water, she gives birth to live babies, and she does so in a way that sometimes scares fishermen. Suddenly the big fish leaps from the water and soars away, looking rather like a monster bat. In the air, she squeezes the young from her body. One by one, they spread their wings, glide down, and begin to feed on particles in the water.

Whales also give birth to live babies. A baby blue whale may weigh as much as two tons when it is born. Its mother pushes it up to the surface for its first breath of air. Then it is ready to nurse. The mother whale's breasts are hidden beneath flaps of skin on her belly. But her baby does not seem to suck, the way most mammal babies do. Instead, the mother pumps milk into its mouth. Nursing can go on only for a few minutes at a time, for the baby must go up for air quite often. Whale milk is the richest kind of food—almost one-third fat. So the baby grows very fast.

Whale mothers watch over their babies carefully. Several times, some whale watchers have seen mothers make surprising rescues of youngsters in trouble. Once, a young whale had raced carelessly toward shore and got stuck on a sand bar. Its mother and another female swam up on either side of it and stopped, with their own bodies partly out on the sand. Then, holding it between them, they heaved and waggled the baby and themselves back into the water.

The white bumps on these grey whales are barnacles.

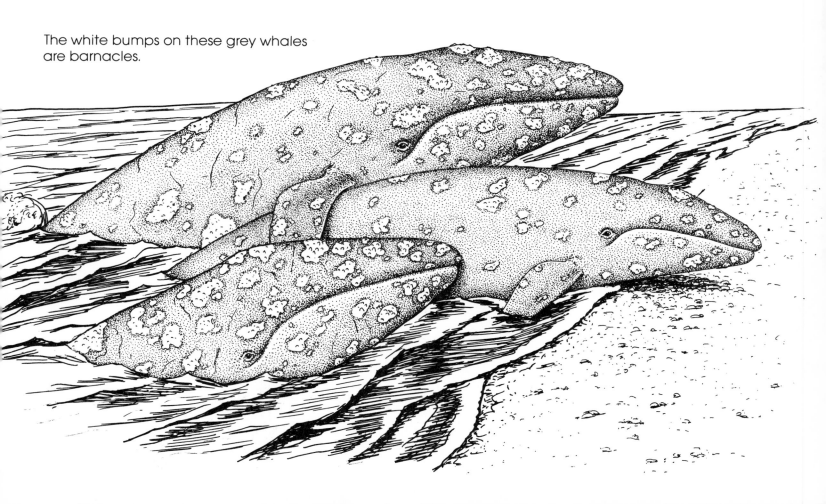

Eerie Lights in the Dark

One afternoon, a young scientist was studying the creatures that live in the sea on the coast of England. He collected a few specimens and then discovered it was too late for him to go all the way home. So he decided to spend the night at an inn. The inn was old and dark, and it reminded him of stories about haunted houses.

Before he went to bed, he carefully set some bottles on the mantel above the fireplace in his room. The bottles held sea water and animals he had collected.

In the middle of the night, some noisy people next door woke the young man, whose name was Alister Hardy. He opened his eyes, and what he saw made him gasp. Ghostly lights were dancing up and down near the foot of his bed. "Like goblins!" he said. "Little blue devils!"

The dancing devils, Hardy soon found, were really very small ctenophores, or sea gooseberries, swimming around in his bottles of sea water. Each of the berry-shaped animals has a special way of producing light inside its body. The light shines through its transparent skin and makes it glow like a tiny lamp. The light-making process is called *bioluminescence*.

Alister Hardy went on to study many other sea animals that can turn on lights in the dark. One kind is less than half as big as the *o* in its scientific name—*Noctiluca*, which means night-shiner. At times Noctiluca will multiply very quickly. There can be three million or more of the tiny creatures in a quart of sea water. Billions of them float about together, shining their lights so that the ocean looks as if it is on fire.

Noctiluca and many other inhabitants of the plankton are called *flagellates*. The name comes from a word that means whip. Their bodies do have a sort of whiplike tail that waggles and moves them through the water. Some flagellates, are animals, and some are plants. Others are a combination of plant and animal. They can make food with the help of sunlight, as plants do, and, like animals, they can also take particles of food out of the water. One kind of flagellate sometimes appears in the Pacific Ocean near the West Coast of the United States. It multiplies quickly and creates a ghostly, shining film on the water. Long ago, whenever Indians saw the glow, they knew it meant trouble. The shining water seemed to be a mysterious signal. It told them not to eat one of their favorite foods—the mussels that lived along the shore. Anyone who disobeyed the signal would get sick and might die. The Indians guarded the beaches at shining-water times to warn people of danger.

The large sea gooseberries are surrounded by white glowing notiluca and two kinds of flagellates, all greatly magnified.

We know now that these tiny, glowing flagellates have bodies rather like droplets of clear jelly. When they float near shore, mussels gobble them up. Each droplet contains a bit of a certain chemical that stays in the mussel's body after it eats the flagellates. The chemical does the mussel no harm. But when human beings eat a lot of mussels, there is enough of that chemical in a shellfish dinner to poison them. No wonder the Indians feared the shining water.

These light-making, poisonous flagellates have a speck of red in their bodies. Vast numbers of them floating together make up what people call "red tides." Soon after the tides appear along a seacoast, the waves bring in tons of dead fish. Unlike mussels, fish that eat the flagellates of the red tide soon die.

The light that makes these living things glow is produced by two chemicals in their bodies. One is called *luciferase*, the other *luciferin*. Neither of them shines by itself. But mixed together, along with oxygen and another special substance, they suddenly light up.

Some sea animals only glow if they are touched or disturbed. This amazed sailors on a British ship named the *Challenger* a little more than a hundred years ago. The *Challenger* was sailing on a long voyage while scientists aboard studied animals that they collected in nets and traps. This was the first time that a whole group of experts ever worked together on a big expedition to discover the secrets of the deep sea. One of the strange things the *Challenger's* nets brought up was a weird looking thing about four feet long and ten inches across and shaped like a tube sock—open at one end and closed at the other. But instead of being one big animal, this was really a whole colony of small ones, hundreds of them, living attached together as if they were a single creature.

On the deck of the *Challenger* the giant, tube-shaped colony lay in a tub, waiting to be examined by the ship's scientists. That night one of the men reached in and touched it. A little blue-green light suddenly appeared. Moving his finger along its side, he traced his name. In a few seconds the named blazed out, he said, "in letters of fire." That is why the animal is called *Pyrosoma*, which means fiery-body, although its fire is not hot.

Many creatures live so far down in the sea that sunlight never reaches them. But they do not live in complete darkness. Tiny sparks flash on and off. Little twinkles glide here and there. Some of the moving lights are carried by fish called anglers. A human angler is someone who catches fish with a pole and line and bait. That is what the deep-sea angler does, too. Its pole and line are a

Angler fish with luminous lure

part of its body that has grown out, long and slender. At the tip of the line a glowing bit of flesh serves as bait to lure the angler's prey.

One kind of anglerfish dangles its bait just above its enormous mouth. Coaxed by the glow, a shrimp or other animal comes close, and the angler gulps it down. In the endless dark, food is not always easy to find. So the angler must grab at anything, even a fish that seems too big to be swallowed. That problem is solved by a stretchable stomach. One angler that a scientist examined had in its bulging stomach another fish three times its own length.

Some fish have glowing lures inside their mouths, but that was not the first thing that scientists noticed when a strange creature was caught not long ago. The men who hauled it out of the water were sure it was a shark. But this was a kind of shark no one had ever seen before. Scientists named it Megamouth, which means big mouth. There were more than four hundred small teeth inside the mouth and light-producing spots as well. The light organs in its mouth, indicated that it lived far down in dark water where the glow attracted prey. A big mystery about Megamouth is this: Why hadn't anybody ever caught one before? There must be more of them in the ocean. Perhaps before long some fisherman will bring another one in.

Some deep-sea animals have a kind of lighting equipment that resembles a flashlight or a lantern. There is a lens in front and a reflector at the back which increases the brightness of the glow. These organs are known as *photophores*. One kind of fish has two rows of photophores on its body. In the upper row are blue, violet and red lights. The lower ones are red and orange, plus a red taillight. Some squids have changeable lights that they switch from red to white and back.

One kind of flashlight fish can blink its lights. Its photophores have a moveable cover—a flap of skin something like an eyelid. Since the photophores lie just below the fish's eyes, it seems to be winking as it swims along. The light in the photophores is not created by chemicals in the fish's own body. Instead, it comes from glowing bacteria that live inside the photophore.

Why do these winking fish wink? No one knows. And scientists aren't sure how the bacteria get into the photophores. Perhaps as the baby fish grow they pick up bacteria from the water around them. Or the mother fish may pass along some of her own bacteria to her eggs.

Many of the other puzzles about animal lights have been solved. Take the case of the little green glowing circles that appeared above certain coral reefs two or three nights after the full moon. Why did some of the circles glow for a while, then disappear, then begin to shine again? What made a circle suddenly get bigger? Where did the lights come from, and why just after the full moon?

It was easy enough to discover that small animals called fireworms flashed the green lights. Their timing was harder to figure out. Scientists now know that the bodies of most living things have a sort of built-in clock. In our bodies, the clocks regulate such things as sleeping and waking. In fireworms, the clock gives the signal for egg laying. And the light of the full moon seems to make the clock's alarm go off.

In the moonlight, a female fireworm swims round and round near the water's surface. Soon her flashing light attracts a male, who fertilizes her eggs as she lays them. If no male should answer her flashing light, the female turns it off for a while, then on again. Sometimes more than one male will come to a female, all of them flashing their own lights. That is what makes the large circles—the fireworms swim around together, laying and fertilizing the eggs. Finally, they all switch off their lights. The mating is over until another full moon.

Hatchet fish   Angler fish   Five-lined constellation fish

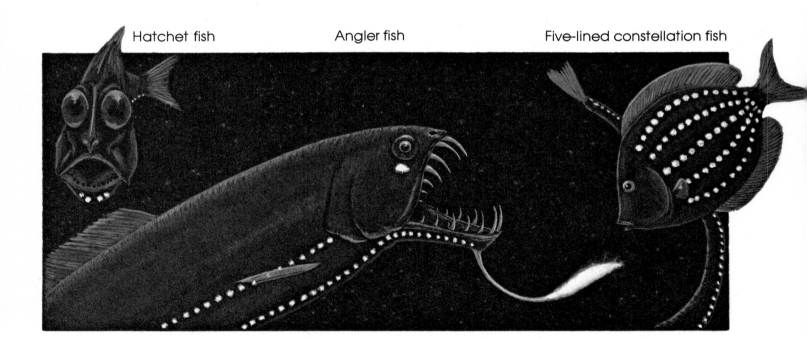

Deep-sea luminescent squids

Scientists believe that a creature's lights are almost always useful to it in some way. But there are still some puzzling cases. Why do bacteria glow all the time instead of flashing on and off? The red lights are another mystery, because most deep-sea creatures are color-blind. That is, like color-blind people, they can't tell red from green. So what advantage can a red light give the animals that carry them?

Various animals seem to flash their lights for protection against enemies. A very bright flash may blind one fish that is pursuing another. When certain shrimps are in danger, they squirt out little bursts of blue light. The lights form a cloud, and the shrimps dash to safety behind it. There are deep-sea squids that defend themselves with clouds of glowing ink. Their pursuers attack the ink while the squids escape. Some worms shed glowing scales. They have time to creep away while the attacker is busy with inedible scales. Other worms have tailights. If an attacker bites the worm in two, the shining tail will most likely be the part that is eaten. Meanwhile, the head part can wiggle away, and the worm will be safe because it can grow a whole new tail end.

How can these worms and some other animals grow new tails or arms or legs to take the place of the lost parts? Why can't people replace lost fingers or hands or feet? That is one of the great mysteries. Scientists hope to solve it before too long.

55

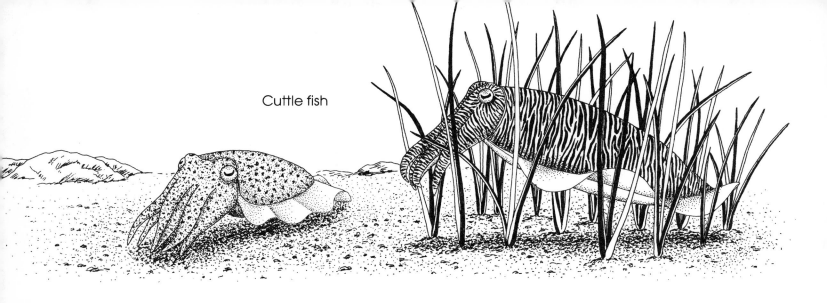
Cuttle fish

Safety Devices

Most creatures have special ways of staying alive so they can reproduce. Some protect themselves by becoming almost invisible. The octopus can change color to match its surroundings. If it is lying on some red coral, it turns red. But, creeping on a sandy sea floor, it becomes the color of the sand. The secret of these quick changes is in little bags that surround different-colored substances in the octopus' skin. Attached to the bags are muscles that act like drawstrings to open and close each one. Bags of red color open and the octopus matches the coral reef. When bags of red and yellow open at the same time, they give the skin an orange look that blends with orange rocks. It takes less than a second for the octopus to put on a new disguise.

The cuttlefish, a cousin of the octopus, will be dark brown one moment, then striped or polkadotted the next. This shifting pattern resembles changing patterns of light and shadow in the water, and it helps to keep the cuttlefish hidden. Scientists know that the muscles that open and shut the bags of color are connected by nerves to the animals' brains. Something in the brain gives just the right signals for color combinations to match the background. How that happens isn't clear.

Small creatures called *nudibranchs* or sea hares change color as they move from deep water toward the shore to lay eggs. In the deep water, rosy-looking young ones live among the rosy weeds they eat. Closer to land, the weeds are brownish, and the sea hares become brown. And so, in both places, they are hard for an enemy to see.

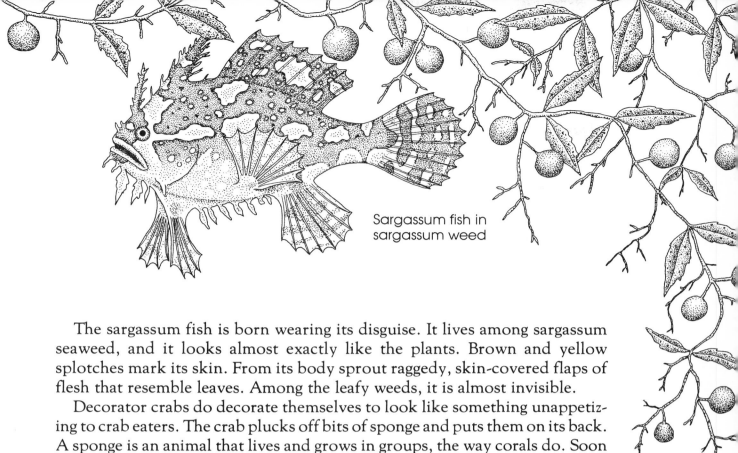

Sargassum fish in sargassum weed

The sargassum fish is born wearing its disguise. It lives among sargassum seaweed, and it looks almost exactly like the plants. Brown and yellow splotches mark its skin. From its body sprout raggedy, skin-covered flaps of flesh that resemble leaves. Among the leafy weeds, it is almost invisible.

Decorator crabs do decorate themselves to look like something unappetizing to crab eaters. The crab plucks off bits of sponge and puts them on its back. A sponge is an animal that lives and grows in groups, the way corals do. Soon the sponge multiplies and covers the crab's back, hiding it from enemies.

The hermit crab is born without a hard, protective covering over its body. But it is far from helpless. It hunts for the empty shells of snails or other creatures and backs its body into them. With its soft parts sheltered and its legs sticking out of its borrowed house, the crab is safe until it outgrows the shell. Then it has to find a bigger one. Often, a hermit crab that lives in a snail shell gives itself extra protection. It disguises itself as a sea anemone, which does not have many enemies because of its stinging tentacles. First, without getting stung, the crab gently strokes an anemone's tentacles. This makes the anemone relax the muscles that keep its base fastened to a rock. Soon the crab has coaxed its prickly new partner to settle down on top of the borrowed shell. From then on, the two stay together and even share food.

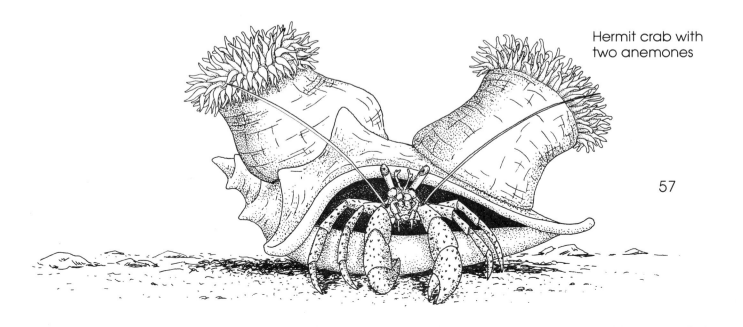

Hermit crab with two anemones

Some bright-colored shrimps and fish seem to get protection because they are valuable to big fish—but not as food. These creatures spend their time cleaning the pests from the bodies and even from the mouths and teeth of big fish. Many fish have these pests—creatures called *parasites*—that live on them, annoy them, and even make them sick. The cleaning shrimps and fish station themselves in certain places where they are easily seen. Big fish come to the cleaning stations and wait their turn to have the parasites removed. The animals that do the work almost never get swallowed by their patients.

The surgeonfish has a spike, as sharp as a surgeon's knife, on its back near the tail. Most of the time the spike lies hidden in a groove. But the fish can whip it out if an enemy attacks.

A group called the scorpionfish are equipped with poison rather like a snake's venom. One of them, the lionfish, carries its poison in sacs near the bottom of some spines on its body. It is such a dangerous-looking creature that divers wonder whether it really needs the poison for protection. Anyway, divers avoid it, and they are also careful not to step on its relative, the stonefish. There are thirteen sharp spines on the stonefish's back. If they pierce a diver's foot they give the most painful sting in the world. If the venom gets into a blood vessel, it can cause death.

Sea snakes, bristle worms, and many other creatures also protect themselves with juices that harm or annoy an attacker. If you touch a fire coral, for example, it is like touching a stinging nettle.

Stingrays carry venom in their whiplike tails. Their cousins, the electric rays, have an electric organ that looks like the sole of a sneaker. It can deliver a shock strong enough to stun an enemy. It can shock a swimmer, too.

Some animals protect themselves in other ways. One kind of sponge keeps attackers away by producing a bad smell. Its name is stinker sponge. Only the angelfish will sometimes nibble it. An animal called the sea cucumber is hunted both by other animals and by people who like to eat it. One member of its family has long, sticky threads that look like spaghetti attached to its breathing organ. The sea cucumber can shoot the threads out with such force that they tangle up an enemy. If threads get into a diver's hair, they can't be washed out, and he may have to shave his head. Divers sometimes use the sticky stuff to bandage a cut.

Another kind of sea cucumber simply shoots out its whole insides when it is attacked. This may give the enemy something to eat, or scare it away. The cucumber now goes to work and grows a complete new set of organs inside itself.

A long sea serpent swims near the smaller worm-like wrasse as they rid a Chinaman fish of parasites. Below are the poisonous lionfish and the blue-tentacled sea cucumber.

Although fish have no eyelids and can't close their eyes, some of them seem to sleep, and that can be a dangerous time. The little coral reef fish, called *wrasse*, that clean big fish sleep motionless in the water. The rainbow parrot fish uses its own mucus to spin a sleeping bag round itself. Now the moray eel, which hunts with its nose, not its eyes, cannot smell the fish wrapped in its bag. The moray must look for other food. One kind of wrasse also sleeps in a mucus blanket.

Flying fish escape enemies by swimming quickly enough to become airborne. They don't truly fly the way birds do. Instead, they beat their tails very fast and get up such speed that they can leave the water and rise on fins, spread out like the wings of a sailplane. Sometimes flying fish glide onto the deck of a ship.

Squids use jet propulsion to escape danger. Some of them squirt water from their siphons with such force that they blast out into the air. Whole bunches of little flying squids may come down in a fisherman's boat or on the beach. One big kind of flying squid moves with such force that it can plow a trench thirty feet long when it lands in the sand.

All these marvellous devices and disguises help to keep the sea world in balance. In daytime, the little wrasse work at their cleaning stations. At night, when the moray and other prowlers are out, the wrasse sleeps safe in its blanket. So the pursued often escapes, while the pursuer seldom goes hungry.

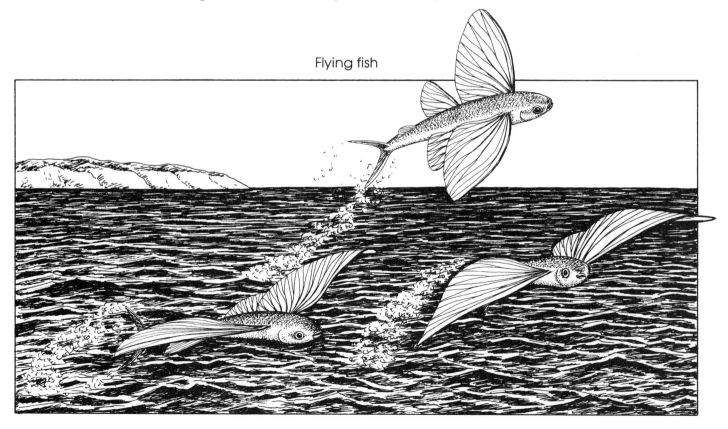

Flying fish

Giant squid

The Monster and the Giant

Ever since he could remember, twelve-year-old Tom Piccot had been hearing stories about a monster called the *kraken*. Sailors said its awful tentacles could coil around a ship and drag it to the bottom of the sea. Most people believed this sea devil existed, but nobody had ever photographed it or brought one ashore to be examined.

Tom Piccot wasn't sure what he thought, until one day in 1873 when he and his father and another man were fishing for herring along the coast of Canada. Not far from their boat they saw something in the water. Perhaps it was part of a wrecked ship. They rowed closer, and one of the men hit the big thing with a boathook. A moment later, a huge head with enormous green eyes rose up. A great mouth opened and a heavy beak bit at the side of the boat. Then a fat arm and a long, thin tentacle grabbed the boat. The kraken!

Slowly the creature began to sink, taking the boat down with it. The men were too scared to move. But Tom reached for a hatchet in the bottom of the boat and chopped off the tentacle and then the thicker arm. As the boat floated free, the monster squirted out a cloud of dark ink, then slithered away.

When young Tom got home he gave the arm to the dogs to eat, but he kept the piece of tentacle. The piece was nineteen feet long, and the whole thing must have been much longer. Whatever this kraken was, it resembled a squid.

Soon after Tom's adventure, more of the monsters turned up. One was caught in a fishing net. Others were washed up on beaches. Now there was no doubt. The kraken was a giant squid, but not nearly so powerful as old-time sailors said it was. It might sink a small fishing boat, but not a sailing ship.

Sperm whales often eat giant squid, which can be more than 60 feet long. But a squid isn't easily taken. It clamps its suckers onto the whale's body and struggles to escape. Scars made by the suckers show on many whales' skins.

Another huge sea creature also has tentacles, but it is not related to the squid. It is the giant jellyfish. It can weigh as much as a ton, and its tentacles are about one hundred feet long. If one of these jellyfish is caught in an icy sea it may freeze and seem to be dead, but it can come to life again after being frozen for hours.

Some people call the great white shark a monster. Others don't agree. They say that movies and stories have made it seem more dangerous than it really is. Nevertheless, scientists take great care when they study these sharks—usually from inside a metal cage with the sharks safely outside. They believe that the animals don't like the way human beings taste, but it is safer not to tempt them to take a bite. The largest great white shark found so far measured about twenty-four feet from nose to tail.

A diver remains safe inside cage while observing great white shark.

A diver gets a free ride from a friendly manta ray.

The largest giant Pacific manta ray was about twenty-four feet from wing tip to wing tip. Usually this big, flat fish avoids people. But once, a young girl found one that was different. She was scuba diving from a boat when she saw a manta ray with a big wound in one side of its head. It must have run into a fisherman's net, for bits of broken rope were caught in its flesh. When the girl came nearer, the fish didn't move. She reached over and gently pulled away some of the rope. The manta ray only shivered. After she had cleaned out its wound, the fish slowly moved and she held on tight. Soon the fish was flying through the water, giving her a ride. After that, it came to the boat every day and took all the members of her family on trips. And there are photographs to prove that all this really happened.

Still other outsize creatures live in the sea. The spider crab measures twelve feet from claw to claw. The giant oar fish grows to be twenty feet long, and is shaped like a ribbon—only twelve inches high from back to belly. It has a mouth that can stretch out to enormous size. When the mouth opens, water rushes in, and the fish's prey is sucked in and swallowed.

The giant sunfish, eight or ten feet long and almost that broad, weighs about two thousand pounds. It is practically round, with almost no tail at all, so it swims by waggling the fins on its back and belly.

Some people believe there must be giant eels somewhere. They think so because somebody once caught an eel that was obviously very young, but it was five feet long. All the other known eels as young as this one are only a few inches long. If there are more such giant baby eels, can it be that they will grow into giant adults, many times as big as ordinary eels? No one knows.

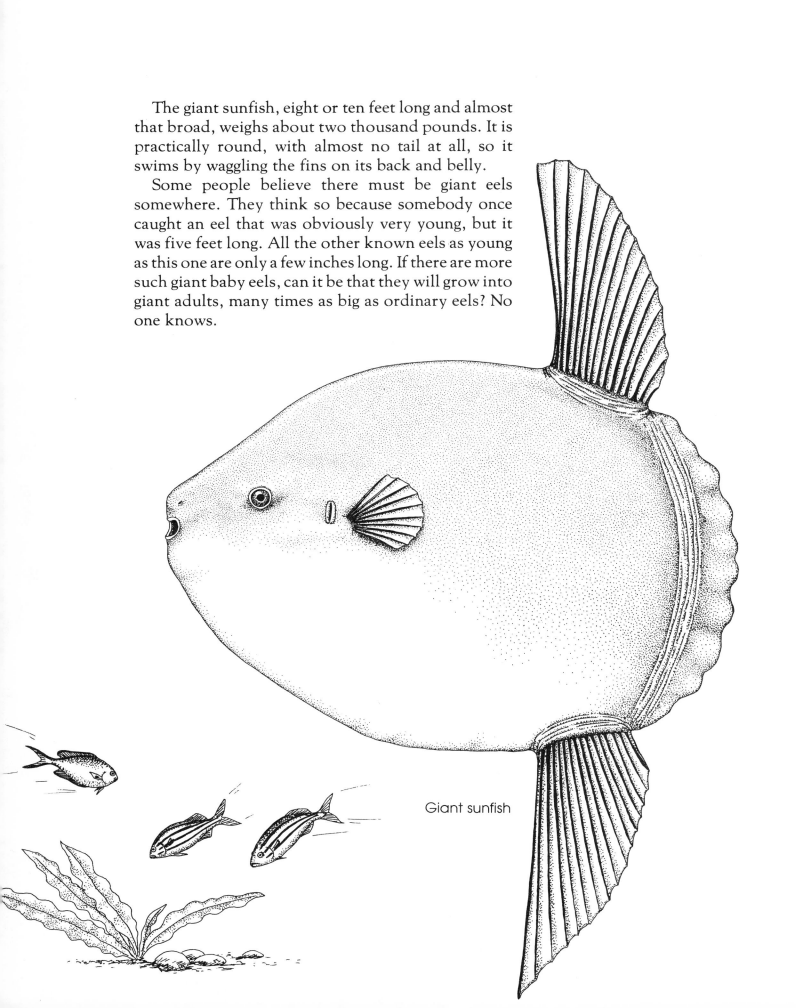

Giant sunfish

Strange Rivers and Perilous Waves

"What is wrong with the captains of the ships that carry our mail?" some English officials grumbled a little more than two hundred years ago. "It takes our ships two weeks longer to sail from London to Philadelphia than it takes them to come back."

Benjamin Franklin listened to the Englishmen's complaints. At that time, Franklin was the postmaster in Philadelphia, and he knew the reason for the mysteriously slow mail. He also could tell what should be done about it.

It happened that Franklin's cousin was the captain of a whaling ship. He hunted for whales in the Atlantic Ocean, and on many voyages he had noticed a sort of river in the ocean. It flowed northeast along the coast of North America, then on toward England. The captain knew that this river, which Franklin named the Gulf Stream, was warmer and a darker blue than the ocean on both sides of it. Whales swam along the edges of the strange stream. They did not try to swim against the current.

The Gulf Stream, said the captain, flows so strongly that it gives a boost to ships sailing from America to England. But if a ship sails in the other direction, the current holds it back. So its voyage takes longer—as much as two weeks longer.

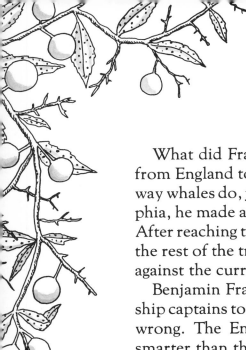

What did Franklin's cousin do to outwit the current when he was sailing from England to America? He traveled along the edge of the Gulf Stream, the way whales do, just outside the current. Then, when he was opposite Philadelphia, he made a dash across the Gulf Stream, which was not very wide there. After reaching the other side, where the current didn't hold him back, he made the rest of the trip quickly. Instead of sailing more than three thousand miles against the current, he struggled through it for only a short distance.

Benjamin Franklin expected that this information would help the English ship captains to get the mail from London to Philadelphia much faster. He was wrong. The English officials thought that no American captain could be smarter than their own captains. So they just said "Humph!" And the mail kept on being slow.

The Gulf Stream is only one of the rivers in the oceans of the world. Columbus found another. It flows from east to west, and it helped his ships on their way to the Americas. The water in this river later joins the Gulf Stream. Together the two currents make a great water system that flows in a giant circle, round and round the Atlantic Ocean north of the Equator.

At the center of this huge circle is a part of the ocean called the Sargasso Sea. The water is calm here, and no strong winds blow. In the old days sailors were terrified of the Sargasso Sea. They knew that great clumps of seaweed collect there, and they thought that a ship could get tangled and trapped in the weeds. That was not really a danger, but the calm water and lack of wind did keep sailing vessels moving slowly or sometimes not at all for a while.

Another name for this unusual part of the ocean was Horse Latitudes. Probably sailors called it that because of a story about a ship that was bringing horses to the New World. The unlucky vessel was supposed to have been trapped in the calm sea for a long time. At last the ship's fresh water was almost used up. There was not enough left for both men and animals to drink. So the crew drove the horses overboard into the sea.

Clumps of sargassum weed

Once some scientists were studying fish in a current in the Pacific Ocean. This current flows from east to west. When the scientists dropped some equipment into the water a strange thing happened. The floats that held up the equipment did not move from east to west with the current. Instead, they began to travel in the opposite direction—*against* the current. What was going on here?

The equipment, which hung at the end of a cable deep under the surface, was pulling the floats along. Only a powerful current could cause that. The scientists had discovered a new river in the ocean, one that nobody had ever suspected before. It flows from west to east, not on the surface, but underneath the surface current, and it carries as much water as several thousand Mississippi Rivers.

There are other such deep ocean rivers. They flow more slowly than surface currents. You move about five times as fast as they do when you go for a walk. Some of the surface rivers flow six or seven miles an hour.

A river in the ocean—the words sound strange. How can a vast current flow along as if the water all around it was almost as solid as land? What causes the currents to move?

Scientists are still not sure they have all the answers. Wind is one great force that moves currents. So do differences in the warmth and saltiness of the water. In hot climates sea water evaporates quickly. The salt remains, and so the water there is saltier than before. The more salt a body of water contains, the heavier it is. So it moves downward, and other water moves in to take its place. Meantime, near the North and South poles, heavy, cold water is also sinking. Cold, deep currents begin to drift toward the equator. They grow warmer along the way. Warm water is lighter than cold, so it rises and mixes with the water in surface currents.

This moving and mixing of currents helps to form the world's climates. The deep rivers bring cooler weather to hot places. The surface rivers carry warmth to places that would otherwise be less comfortable. Deep water that rises to the surface brings up from the ocean bottoms food that nourishes creatures in upper levels.

These are only a very few of the things that scientists think about when they wonder how and why the ocean's rivers flow. They have been studying and measuring the temperature of sea water for more than a hundred years. Sometimes the job was frustrating. When they lowered a thermometer in a glass bottle wrapped in a bundle of soft cloth, what came up? A handful of fine powder. The pressure of water deep down squeezed the glass into particles smaller than sand. Later, thermometers that don't get crushed were invented.

One man who wanted to find out more about the ocean was a very rich prince named Albert in the tiny country of Monaco. In the year 1885 he used some of his money to investigate currents. On one expedition, he tossed 2000 bottles overboard into the Atlantic Ocean. Each one contained a message in nine languages asking the finder to report where it was picked up.

The bottles floated and drifted along with surface currents. Many of them were found. As the reports came in, Albert marked the bottles' locations on a map. In the end he could show where some of the ocean's rivers flowed.

Today, very complicated instruments give scientists much information about currents. So do satellite photographs, taken by cameras in space. But scientists still rely on bottles for a great deal of information. Each year they launch about 20,000 drift bottles. A surprising number are found and reported.

Like ocean currents, ocean waves are not yet completely understood. Everybody agrees that the ordinary waves you see at the shore have been started by winds. But exactly how this happens no one can say for sure. One thing we do know. Several small waves may join together somehow and form a single very big one. Sometimes these gigantic waves rise up unexpectedly and strike ships as sea. The tallest one ever measured accurately was 119 feet high. When it crashed down, it was as if a green wall, ten stories high, was toppling onto the ship's deck.

Another kind of wave has enormous power, but ships on the surface of the ocean hardly notice its mysterious passage. This kind is called an internal wave. It disturbs only the deep water. What causes an internal wave? Why doesn't it disturb the surface? No one has yet discovered the anwers to these questions.

A *tsunami*, which used to be called a tidal wave, starts with an earthquake or a vast landslide deep under the ocean. The shock sets a great body of water into motion, and a wave begins to travel outward. As it approaches land, a surprising thing happens. The water near the shore flows away as if it is being sucked outward. Then, with a terrible roar, the sea returns. A mountain of water falls on the land, and almost everything beneath it is crushed. Tsunamis have destroyed whole seaport towns and killed thousands of people. Carried on top of a giant wave, large ships have been set down a mile or more inland.

Scientists now have instruments to detect earthquakes, and they can send out warnings that a tsunami is coming.

Household sponge

Solving Mysteries— Preserving Life

On almost any day, more than two thousand years ago, Greek divers came home with large floppy chunks of stuff in their boats. Soon the leathery outside of each chunk rotted away, leaving a soft spongy skeleton. This was, indeed, a sponge—a real one. It had been growing on the bottom of the Mediterranean Sea until a diver pulled it loose and brought it up to his boat.

No one knows exactly when or how people around the Mediterranean Sea discovered sponges. But the dried and cleaned sponge skeletons had many uses. Mothers soaked a bit of the soft stuff in honey and let babies suck on it—a good way to stop them from crying. Doctors made sponge bandages for cuts and wounds. In those days, soldiers wore heavy, uncomfortable metal helmets. But a soft sponge on top of the head kept the helmet from hurting. Of course, people also used real sponges as we use plastic ones today—for washing floors or dishes or themselves.

People who live a long way from the sea sometimes have a disease called goiter that most of those who live near the sea do not get. Doctors long ago discovered that they could treat this disease if they burned sponges and gave the ashes to their patients. They did not know why this helped. Later, scientists discovered that sponges and sea salt contain iodine, a chemical that people's bodies need. Those who lived near the sea got enough of it to prevent the disease. Burned sponges provided it for others. Today our salt often has iodine added to it at the factory where it is packaged.

There are many kinds of sponges.

Sea squirts

Barnacles

The search for medicines from the sea still goes on. Perhaps new antibiotics can be made from sponges or other sea animals. One researcher hopes to cure colds and flu with a drug made from a chemical found in creatures called sea squirts. A scientist who was studying the poison darts in the Portuguese man-of-war made a discovery that has helped people who have very bad allergies. Perhaps some of the poisons that protect other sea animals will turn out to be useful drugs. Dentists would like to know more about the glue that the decorator crab uses when it sticks bits of sponge on its back for disguise. They have already glued false teeth in place with the sticky stuff that holds barnacles tight to their rocks.

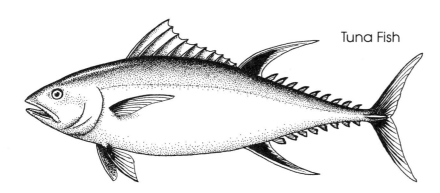
Tuna Fish

The sea has always been a source of food, but the work of getting it is hard. Instead of making long trips in fishing boats, why not build pens near shore where fish can be raised? Japanese fish farmers do just that. They produce tons of yellowfin tuna every year, and there are fish farms in other countries, too.

A seaweed called kelp grows along the seacoast in many places. It feeds and shelters crowds of fish and other creatures. People harvest it to make food for chickens and farm animals. Kelp also contains a sticky stuff that food factories use. It makes ice cream and chocolate milk especially creamy. It gives fluffiness to imitation whipped cream. Another kind of seaweed provides a substance that thickens sour cream and puts the jell in jellybeans.

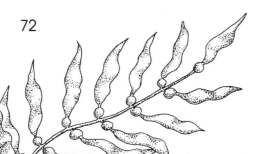
Kelp and other seaweed

If you have been to the seashore, you have probably seen the water creeping steadily up onto the beach. Then, at its highest point—high tide—it stops and begins to retreat. Tides are vast sloshings of the ocean, back and forth, in rhythm with the moon as it travels around the earth. As if it were a magnet, the moon pulls on the ocean and draws the water up into a hump. When the hump is rising, water leaves the beaches. After the moon has passed by, the hump flattens. The water spreads out, and the tide comes in again.

Strong tides can force their way up rivers as well as onto beaches. Long ago people began putting this force to good use. They built special dams in rivers and trapped the water that came in with the high tide. As the water flowed out again, it turned wheels in mills that ground grain or produced power for machinery.

Electric power now runs most of our mills and factories. But the idea of using tide power is still alive. In France and the Soviet Union, dammed tide water now produces electricity. People in other countries are thinking about building that kind of power station.

The deep-sea explorer, *Challenger*

Much of our electricity comes from power plants that burn oil. The oil is millions of years old, but where did it come from? The sea. Long before there were people or even dinosaurs on earth, tiny plants, such as diatoms, and microscopic animals lived in the oceans. Trillions and trillions of them multiplied and died and dropped to the sea floor. Their skeletons and shells piled up in layer after layer that grew heavier and heavier and finally hardened into rock. Pressure of the rock, together with heat from inside the earth, turned the remains of plants and animal bodies into oil. Perhaps bacteria helped to make the oil, too. This ancient oil has collected in pools deep under the earth. Recently scientists have discovered new oil being made by pressure and bacteria near hot spots in the sea floor. It is different from the oil that comes from oil wells—but then it is not millions of years old.

In some places diatoms were almost the only living thing that covered the ancient sea floor in deep layers. Later those places were pushed upward and became dry land. Finally people made their homes near the patches of those very old diatom shells. They saw that pressure had broken the glassy shells into tiny sharp fragments, held together in a creamy white mass. Was there any use for this unusual stuff? The answer was "yes." Power shovels now dig up quantities of it to be made into scouring powder and insulation. The bright little glass needles, added to paint, make the white lines on roads shine and reflect a car's headlights at night. The black asphalt for the road itself came from some of the millions-of-years-old sea life. Diatoms have given drivers both oil that is used in making asphalt and paint for the safety lines that mark it.

Sea water is full of minerals washed down into it from the land and spilled up through hot springs and underwater volcanoes. One of these minerals is salt which people harvest by letting sea water evaporate. Silver, copper, nickel, and other valuable minerals are also dissolved in the water. One called manganese collects in hard chunks called nodules on the sea floor. Manganese is used in making steel. It now has to be dug out of mines in the earth, but it could also be mined from the sea.

Mining companies all over the world would like to get minerals from the sea. But there are big questions that have to be answered first. Can anyone own a piece of the bottom of the sea? Can any country claim a piece? If so, who decides which country has a right to mine certain places?

Many important people believe that the sea belongs to everyone. No person and no country should be allowed to say, "Stay off. This undersea area belongs to us." Some of these people have written what they call the Law of the Sea. The law would make sure that mining and other activities do not destroy sea life or give any group of people more than its share of the sea's riches. Some countries have agreed to the law. Some countries have not. What has your country done about it?

Scientists especially want everyone to help protect the sea and sea life. Tankers sometimes spill oil in the water. Or an accident at an undersea well may let millions of gallons of oil escape. Much of it floats to shore and spoils beaches. On the way it kills fish. It coats the feathers of birds so that they cannot fly. Many of them die. When garbage or radioactive material or chemicals are dumped, even in deep water, they do not just disappear without doing any harm. These foreign things disturb the order and rhythm of nature in the oceans. But preserving this order and rhythm helps to preserve our own life on land.

Solving mysteries of the sea is a great step toward preservation. It is also great fun.

Glossary

Bacteria (back-TEE-ree-yah)   Very simple, tiny plants. Each one is just a single unit, called a cell. It has no roots, stems, or leaves, and it does not contain the substance called chlorophyll that green plants have.

Baleen (buh-LEEN)   In the mouths of whales that have no teeth, strips of a yellowish substance called baleen hang down from the upper jaw. Baleen is made of the same substance as fingernails, and the strips are called plates. Fringes on the plates act like sieves to strain from seawater the small creatures that these whales eat.

Bioluminescent (By-oh-LOOM-in-ess-ent)   Some bacteria, fish, and other animals can produce light in certain parts of their bodies. They are called bioluminescent. The word comes from two words that mean *living* and *light*.

Carbon dioxide (CAR-bun die-OX-ide)   A colorless gas that can dissolve in water. Sea plants need it, along with other substances and sunlight, in order to grow.

Chlorophyll (KLORE-oh-fill)   A green coloring substance in plants. It takes in sunlight and gives plants the energy they need in order to grow. See *photosynthesis* on this page.

Cilia (SILLY-ah)   Delicate, threadlike, microscopic parts that grow on or in the bodies of animals. By waving back and forth in a steady rhythm they move one-celled sea animals through the water. For larger animals they create currents that sweep food toward the mouth or waste material out of the body. See *flagella* on this page.

Ciliates (SILL-ee-ates)   A group of animals that have cilia.

Crystalline (KRIS-tal-een)   Scientists use this word to describe the clear, hard skeletons of diatoms. Unlike glass, which is formed by heating a mineral called silicon, the crystalline skeletons are made by the diatoms from silicon dissolved in sea water.

Flagella (fla-JELL-ah)   This word comes from a Latin word that means *whips*. Some sea animals and some sea plants have long, thin waving parts called flagella that whip about and drive them through the water.

Flagellate (FLA-jell-ate)   Animals and plants that have flagella are called flagellates.

Gill rakers   Some fish have long, needlelike bones that form a sort of curtain in their mouths in front of the opening where water is squeezed out through their gills. These curtains are called gill rakers. They keep bits of food from clogging up the gills.

Larva (LAR-vuh)   Just after hatching from an egg, a very young sea animal often does not look at all like its parents. It is called a larva.

Luciferase (loo-SIF-er-ase) and **Luciferin** (loo-SIF-er-in)   When these two substances combine with oxygen, they produce the light that makes glowing spots on the bodies of certain animals. See *bioluminescent* on this page.

Parasite (PARE-uh-site)   Some animals and plants live on or inside another living thing and get their food from its body. They are parasites.

Photosynthesis (FOE-toe-SIN-the-sis)   This word comes from two words that mean *light* and *putting together*. When a plant builds its body by putting together carbon dioxide and water, with the help of sunlight and a green substance called chlorophyll, the process is called photosynthesis. See *chlorophyll* on this page.

Photophore (FOE-toe-fore)   Some fish and other sea creatures have parts of their body where light is produced. These spots on the skin are called photophores.

Plankton (PLANK-ton)   Many different kinds of very small plant and animal live in the sea, usually rather close to the surface. All of them together are called the plankton. The plants of the plankton are called phytoplankton (FIE-toe-PLANK-ton). The animals are called zooplankton (ZOE-OH-PLANK-ton).

Tentacles (TEN-tuh-kuls)   The long, thin, movable armlike parts of certain animals. Some are used to capture prey, some for swimming, feeling, clutching, or for other purposes.

Index

Alvin, 16, 24
Anemone, 20, 21, 22, 57
Anglerfish, 52-53

Bacteria, 18, 53, 54, 55, 74, 76
Baleen plates, 31, 32, 76
Barnacles, 27, 30, 44, 48, 72
Bioluminescence, 50-55, 76
Bluefin tuna, 12, 38
Blue marlin, 47
Blue whales, 49
Bottles, 68
Bristle worms, 58

Calcium carbonate, 22
Carbon dioxide, 10, 76
Cartilage, 37
Challenger, 10, 52
Chlorophyll, 10, 76
Cilia, 26, 27, 76
Ciliate, 27, 76
Clams, 16, 17, 19, 26, 28
Coccolith, 14, 15
Codfish, 30, 37
Colonial animals, 22, 24, 25, 52,
Corals, 22, 40, 54, 56, 58
Crabs, 16, 30, 47, 57
Croaker, 39
Ctenophore, 12, 50
Currents, 65-69
Cuttlefish, 56

Damselfish, 40
"Dandelion," 16, 17, 24-25
Dead Man's Fingers, 22
Decorator crabs, 57, 72
Diatoms, 9, 10, 12, 74
Dolphins, 38, 43, 44

Ears, 33
Eels, 46, 64
Eggs, 23, 27, 30, 38, 42, 47, 48, 54
Electric rays, 58
Electric sense, 37
Elvers, 46
Eyes, 34-35

Feather star, 28
Fiddler crab, 27

Fire coral, 58
Fireworms, 54
Flagella, 12, 76
Flagellates, 15, 50, 76
Flatworms, 35
Flying fish, 60
Foraminifera, 11, 12, 38

Giant eels, 64
Giant jellyfish, 62
Giant oar fish, 63
Giant squid, 62
Giant sunfish, 64
Gill rakers, 31, 76
Goiter, 70
Greek divers, 70
Gulf Stream, 65-66

Hardy, Alister, 50
Hermit crabs, 57
Herring, 30, 33, 35, 36
Horse Latitudes, 66
Humpback whales, 32, 33, 44-45

Indians, 50, 52
Internal waves, 69

Janthina, 25
Jellyfish, 23-24, 35, 62

Kelp, 72
Killer whale, 32
Kraken, 61
Krill, 12, 13

Larva, 47, 76
Lateral line, 33, 34
Lava, 16
Law of the Sea, 75
Lights, 50-55
Lionfish, 58
Luciferase, 52, 76
Luciferin, 52, 76

Magnetic sense, 38
Mail ships, 65
Manatees, 2-3, 78-79
Manganese, 74
Manta rays, 49, 63
Medusas, 23-24
Megamouth, 53
Mining, 75
Moray eel, 60
Mucus, 60

Mussels, 16, 50, 52

Noctiluca, 50
Nose pits, 37
Nudibranchs, 56

Oar fish, 63
Octopus, 29, 34, 47, 56
Oikopleura, 14-15
Oil, 74, 75
Opossum shrimp, 34, 48
Oxygen, 19
Oyster, 29

Parasites, 58, 76
Parrot fish, 40, 60
Photophores, 53, 76
Picot, Tom, 61-62
Pistol shrimp, 41
Plankton, 12, 15, 76
Pods, 45
Poison darts, 21, 23
Poisonous animals, 24, 25, 52, 58
Pollution, 75
Polyps, 22, 24
Portugese man-of-war, 24, 25, 72
Pyrosoma, 52

Radarlike sense, 37
Radiolaria, 10, 11, 12
Radula, 29
Rainbow parrot fish, 60
Rays, 37, 40
Red tides, 52
Reefs, 22, 54, 56

Salmon, 37, 38
Salps, 12
Sargasso Sea, 46, 66
Sargassum fish, 57
Sargassum weed, 57, 66
Scallops, 34
Schools of fish, 35, 36
Scorpionfish, 58
Sea cucumber, 58
Sea gooseberry, 12, 34, 50, 51
Sea hares, 56
Seahorses, 48
Sea lamprey, 29
Sea slugs, 21
Sea snails, 19, 25, 29, 30, 35, 57
Sea snakes, 58,
Sea squirts, 72
Sea urchins, 30
Sea wasp, 24
Senses, 33-38, 41
Sharks, 30, 36, 37, 53, 62
Shells, 10, 11, 12, 19, 28, 57, 74
Shrimplike animals, 12, 19
Shrimps, 34, 41, 55, 58
Siphons, 26, 29, 60
Sleep, 60
Snails, 19, 25, 29, 30, 35, 57
Soda water, 10
Songs of whales, 45
Sounds, 39-45
Spider crabs, 63
Spiny lobsters, 40
Sponges, 38, 57, 58, 70-72
Starfish, 28, 30, 34
Stingers, 24
Stingrays, 58
Stinker sponge, 58
Stonefish, 58
Submarines, 16, 24, 39
Suckers, 29
Sunfish, 64
Surgeonfish, 58
Swim bladders, 33, 39, 48

Taste buds, 37
Teeth, 29, 30
Tentacles, 20-25, 34, 57, 62, 76
Tides, 73
Toadfish, 42
Tsumanis, 69
Tubefeet, 28
Tuna, 12, 38
Turtles, 38

Velella, 25, 35
Vents, 16, 18
Vitamins, 10

Waves, 69
Whales, 12, 31, 32, 33, 44-45, 49, 62, 65
Whiting, 37
Wolf eel, 30
Worms, 16, 17, 19, 35, 55

Yellowfin tuna, 38

Manatees